ON THE EMERGENCE OF
HUMAN CONSCIOUSNESS

RAFAEL PINTOS-LÓPEZ

CONTENTS

For Inés, who had to put up with
even more lunacy and more obsession.

For Hernán, Rodrigo and Millán,
three high achievers who are
the heroes of my story.

"Here's to the crazy ones. The misfits. The rebels. The troublemakers. The round pegs in the square holes. The ones who see things differently. They're not fond of rules. And they have no respect for the status quo. You can quote them, disagree with them, glorify or vilify them. About the only thing you can't do is ignore them. Because they change things. They push the human race forward. And while some may see them as the crazy ones, we see genius. Because the people who are crazy enough to think they can change the world are the ones who do."

— Steve Jobs, 1997

∽

"Academia is to knowledge what prostitution is to love; close enough on the surface but, to the nonsucker, not exactly the same thing."

— Nassim Nicholas Taleb

～

"...The pioneer scientists who are exploring consciousness and the mind such as Gerry Adelman, Brian Josephson and Francis Crick, what they have in common is they've all won the Nobel Prize and, perhaps after that, it's something you can do; otherwise, people don't take you seriously."

— Susan Greenfield

～

"It seems plain and self-evident, yet it needs to be said: the isolated knowledge obtained by a group of specialists in a narrow field has in itself no value whatsoever, but only in its synthesis with all the rest of knowledge and only inasmuch as it really contributes in this synthesis toward answering the demand, 'Who are we?'"

— Erwin Schrödinger

～

"Our universities fail to guide us down the easiest paths to wisdom... Rather than teaching a sense of awe, they teach the very opposite: counting and measuring over delight, sobriety over enchantment, a rigid hold on scattered individual parts over an affinity for the unified and the whole. These are not

schools of wisdom, after all, but schools of knowledge, though they take for granted that which they cannot teach — the capacity for experience, the capacity for being moved, the Goethean sense of wonderment."

— *Herman Hesse*

PREFACE

"The traits I respect the most are
erudition and the courage to
stand up when half-men are afraid
for their reputation.
Any idiot can be intelligent."
- Nassim Nicholas Taleb

The main aim of this book is to attempt what previous generations would have considered a challenge for a gentleman: to right a wrong, to propose a solution to an insoluble problem: a gentleman should fight even when the cause seems a lost cause.

In this case, what seems to be a lost cause is that the scientific world has been doggedly trying to solve a problem that has no solution; and that is because the question it has posed is the wrong question. However, when science is bent on a method, the answer to alternative methods is to ridicule them. What is not mainstream is considered crackpot. Nobody wants to hear.

Please understand, this is not a fight against universities, or academia, or science. It is an indictment of a mentality that shuns and ridicules what is different.

The emergence of human consciousness does not fit easily within Darwinian evolution. Wallace knew it. Darwin suspected it. There is no physical explanation. The only explanation is dualistic. The non-physical nature of human consciousness cannot be ignored.

The mainstream Neo Darwinian scientific world attempts an explanation of consciousness solely following a gradual accumulation of random mutations through natural selection. That is wrong on two accounts: 1) evolution can also occur through alternative mechanisms like spontaneous mutations, i.e., genetic insertions, or symbiogenesis; 2) consciousness in humans has a cultural component that cannot be explained through physical evolution. The main reason for the scientific bias towards physicalism is that biology is a physical science and no explanation can be found for consciousness within its bounds.

The extreme application of Bayesian logic to the study of consciousness, in current neuroscience and philosophy of science, has resulted in the negation of a reality that should be apparent to everyone. Sometimes, probabilities provide no answer.

Human consciousness is a black swan phenomenon. Aeons of evolution could not have predicted it. It just happened. Nassim Taleb is laughing at the scientific world.

RP-L

INTRODUCTION

"What is your aim in philosophy? -
To show the fly the way out of
the fly-bottle."
- Ludwig Wittgenstein

The title of this book might be misleading. That was not the intention, of course. Unfortunately, many readers will probably bring a common misconception to this encounter. The purpose of the book is to change their mind and show them a miracle happening before their very eyes.

My guess is that the word *"emergence"* brings ideas of a foggy time in the past, when our cave-dwelling ancestors grunted to each other and when they commenced to utter and understand language (I have to confess that the idea is fascinating). But the book is not actually about that; the book is about an ongoing process that takes place around us and that many of us probably do not connect with human consciousness.

The news is this: you and I were not born with our human capacities completely developed. We were born only equipped to be fully human. But we could neither speak nor communicate. We did not understand our parents, our siblings, our relatives. They loved us and cared for us. And we could do nothing but allow them to love us and care for us. With speech, with communication, comes cognition; all of them are skills that humans need in order to live in society. We possess a highly adaptive capacity to learn from our experiences, but this capacity would not have survived prehistory without the cumulative generation-to-generation transmission of knowledge.

Had it not been that way, the human species would not have developed the way it has. Literacy, numeracy and dexterity are acquired, they are not innate. Looking for them in the brain as evolutionary phenomena is an exercise in futility.

The miracle that is happening right now is that many parents, and relatives, and the collective, are teaching countless infants how to speak and communicate. They socialise them and teach them. Those children are acquiring what is necessary to become useful members of human society. That is not only the emergence of individual consciousness, but it is the process that integrates *human* (and thus collective) consciousness to the innate biological component of consciousness that we bring with us to this world. As they acquire cognition, those children are becoming part of the collective consciousness of humanity.

Now, there is something else that needs to be said.

∼

The decision to write this book was, to some degree, absurd. The absurdity did not lie in writing the book itself, which was carried out with the greatest amount of common sense I could muster. Among the quotations on the page before the preface, there is one by Susan Greenfield that sums it all up. The scientific and academic establishment is not interested in very divergent ideas about consciousness. Only after winning the Nobel Prize, you might have the chance of being taken seriously. Otherwise, it is highly improbable. In any case, write it I must.

Apparently, the concept of a two-tiered consciousness —which is pure common sense, a sound notion, and one that would simplify research (if I say so myself) —does not appear to be of any help to current scientific theories, muddled and opaque as they are. Science is at a stage of quasi-religious search and dogma. There is no falsifiability when we discuss the consciousness of particles, or of the universe. Even when that is so, there is no screening for crackpot theories. All theories from non-practising-academics/researchers are treated as crackpot. Researchers and academics prefer their own views and methods, and will tend to ignore any innovations on the subject.

Human knowledge has developed through history on the basis of two groups of individuals: the innovator (the person who had the idea of the wheel) and the educator (the person who taught the following generation about the wheel). Innovators are lateral thinkers; educators and researchers are not, although they may improve ideas through hard slog and experimentation. Both groups are required to achieve a balanced development of knowledge.

I recently corresponded with a renowned physicist who told me that my notions on consciousness were novel, but my deduction about the nature of time was already conventional. I say time is a human construct; actually, a sort of scaffolding or infrastructure where we can place instances of change. In any case, I was very grateful because he had taken the time to read what I had written.

What happened when *Homo Sapiens* began to acquire long-term memory and imagination? What happened when an explanation was needed for past and future, and causality? It would be possible to say that, at that point, we "discovered" time. It would be possible to say that we "accepted" a new, prolonged reality that involved past and future. I'd prefer to say that—unwittingly—we "invented" time as an explanation for causality, long-term memory and imagination.

At one point in history, human beings, I am sure, experienced the strange feeling —much more clearly than other developed animals—that there was something intangible, ineffable, that contained change and allowed it to flow. In the case of humans, something that included the lives of their parents and ancestors, that explained things that had happened generations before. For social reasons, that something—that entity we called time—necessitated measurement. To some cultures, this new notion appeared as a weirdly linear entity, the behaviour of which they could not quite comprehend. The more the species evolved and cultures evolved, the more sophisticated measurements became. Time-measurement as a development within the different cultures is obvious because most of it happened relatively recently, within written history, although the Tanakh, which is based on oral traditions, vaguely mentions periods and ages, prior to becoming Scripture.

∼

But, going back to consciousness, there is also an ontological difference between what current science considers consciousness and what the concept should be if we view it from a different, more logical, perspective.

I question the term "consciousness"—as currently used by science—semantically. Unless we deem human consciousness an integrated whole where sentience and sapience coexist as two discrete entities, it is impossible even to begin to analyse the phenomenon and expect logical results. A new definition of consciousness—one that would include the input of linguists and philosophers—is important and it would make research much clearer.

Somebody recently asked the question: "Is my dog conscious?". The person actually meant: "Is my dog sentient?". Of course, it is. But it is not conscious. The difference is massive and it involves the fact that *Homo Sapiens* has an added layer of consciousness: *psyche*. Other species may even have some degree of cognition. Only humans have metacognition.

Human consciousness, which is what science is most interested in, should involve the possibility of cognition and metacognition, of the individual asking himself/herself/itself, "Why am I feeling this?" If the individual cannot ask this question, then it should not be considered a conscious being. That leaves other mammals and AI out of the equation. Without a sentient body, Artificial Intelligence lacks sensations or feelings, it is not alive; other animal species lack *psyche*; they cannot think, nor can they think about thinking. I suppose when we discuss consciousness, the main objective in which we are all interested is *human* consciousness.

What human individuals have acquired thanks to culture is the possibility of functioning much more efficiently within a plurality. For different reasons, other species have similar capacities. But they cannot transfer complex information. A flock of birds can fly in a formation (called a "murmuration"): individual birds somehow know what to do, and they do it in unison with the rest of the flock. Starlings, swallows, and other birds, can fly simultaneously in large numbers without a leader. They appear to do it effortlessly, and they know when to change direction. So far, the purpose of that type of flying is not quite clear. Some say that it is to avoid birds of prey. Birds also migrate in flocks. Schools of fish behave in similar fashion.

Apes, like chimpanzees, live in groups similar to human clans. To some extent, their behaviour appears to be innate. One fundamental difference with human beings is that, even when they may learn socialisation to some degree, they cannot pass knowledge from one generation to the next one, as they lack language.

The amazing thing about human society is not that its individual members are conscious. It is that society itself is conscious. Science has not found a method to quantify that phenomenon as yet. It remains, probably because of Galileo Galilei—who dictated how knowledge should be divided—a discrete humanistic phenomenon. The dualistic approach proposed by Descartes—currently out of fashion within the scientific world—appears to be the only solution to explain how intangible phenomena like the workings of society and culture can be integrated into a physical entity like the mind of a human individual.

∼

What I propose is that human consciousness includes two components: a basic one and a high one, and that they are discrete, evolved from different needs of our species, and evolved during different periods. I prefer to think of them as layers, as the term implies a more visual way of understanding that the high one is a superimposition on the basic one. Literally—from a physical point of view—because the neocortex is a massive neural addition enveloping the less developed areas of the triune brain.

Humans might not be the only species to have discrete layers (for sentience and some cognition). A recent study on macaques conducted by an international team headed by Sean Froudist-Walsh at the University of Bristol, confirmed that there is a separation between sensory and cognitive networks in other species as well. Brain receptors appear to operate on the basis of patterns that shed light on their function in terms of perception, memory, and emotion.

In any case, studying one type of consciousness will not help solve the problems associated with the other, as they are two different phenomena, discrete and overlapping, not a continuum. This is a matter that appears to complicate the "hard problem of consciousness" as formulated by Chalmers. It provides, however, a clearer path towards a solution to the problem of human consciousness if the problem is expressed in different terms. Of course, human consciousness cannot be explained by reducing it to its physical constituents. Qualia are essential, indispensable. All living beings must have them to survive. That is another problem.

The acquisition of human consciousness took some aleatory turns. Our species arrived at it through our relative physical weakness and our need for a lengthy upbringing. Like some birds, we are an altricial species: we are born extremely immature and require maternal and collective care for many years.

The development of this new type of consciousness was refined and exponentially accentuated as groups of humans became larger and larger. And it continues to develop and become more sophisticated. The more successful and the more gregarious we became, the more human consciousness grew. In the process, it acquired metacognition, elaborate thought, long-term memory, creativity, individual and social identity, and time, among other qualities and concepts.

In our current societies, infants need to be taught language, literacy and numeracy, and they have to be socialised, before they can become full members of the collective. These abilities appear to have increasingly evolved from the moment humans began living in larger groups, thanks to language, trust, and food-sharing, and have indeed transcended standard evolutionary biology.

In summary, I claim as follows:

Human consciousness consists of two integrated but discrete layers:
1) basic animal consciousness (nephesh)
2) high human consciousness (psyche).

* High consciousness is only acquired through parental and collective upbringing. It is individually transmitted. Its nature is cultural.*

* This confirms the Sapir-Whorf hypothesis concerning linguistic relativity, and falsifies universalist claims concerning language, as proposed by Noam Chomsky.*

* Understanding the workings of high consciousness cannot be arrived at through an*

evolutionist study of basic— biological —consciousness.

* Imagination, creativity, language, long-term memory, adventurousness, are exclusively human traits acquired through high conscious-ness, i.e., through culture.

* The logical conclusion is that there should be cortical and other centres in the brain, newer than any centre that deals with strictly biolog-ical phenomena, where cultural developments are processed (e.g., Broca's and Wernicke's areas).

* Time is a human construct that exists only within high consciousness, through unlimited imagination (expectation) and long-term memory (which involves identity and collec-tive perception).

* Without high consciousness there is only present and change.

Let us see how we can explain all these hypotheses.

CHAPTER 1

OVERVIEW

"Life is not a series of gig lamps symmetrically arranged;
life is a luminous halo, a semi-transparent envelope
surrounding us from the beginning of consciousness to the
end."

\- Virginia Woolf

*H*uman consciousness is not sentience. Human consciousness is nothing without society.

As far as I can tell, no one appears to have concrete answers concerning the question of human consciousness. Neuroscientists can observe neurons, brain centres and synapses communicating, and operating in certain ways when human beings imagine, remember, or feel emotion as they listen to music, watch a movie or a painting. But those behaviours are not the product of physical evolution as they want to believe: they are the product of culture. There is no music, or movie, or painting without a culture behind them.

I want to start this introduction illustrating the position of scientists, with a segment of an interview, a question from a journalist to Prof. Susan Greenfield:

> *Radio Journalist*: "... *But what about that general sort of global consciousness that we all experience, what generates that feeling of a sense of self? Connections between neurons suggests Susan Greenfield, but on a much larger and more transient scale*".
>
> *Prof. Greenfield*: *This is the big problem because at first blush it is qualitative and scientists hate qualitative things, so if only we could turn it into something we could measure. The one thing that we could measure would be the intermediate level of these connections but not just the hardwired ones but the ones that can change perhaps, that can corral up very quickly and disband. We know that the brain can do that, that if you flash lights for example at the brain, we know that in a quarter of a second, ten million neurons can start working together and then by half a second they are disbanded again.*"

This is typical. There is no answer. She has not provided one. Whenever you mention human consciousness, whether collective or individual, scientists are deliberately vague or ambiguous. This is what I have been finding they say when asked about information, communication and other issues that relate to how individuals communicate, how society operates from a conscious perspective, how they explain what—in this case—the journalist calls *"global consciousness"*.

The *"need to measure"* that Professor Greenfield mentions has to do, of course, with the philosophical concept of "objective reality". There is a reality that we all share. If you are able to measure it, you can prove to me that what you say is right. That is the way science operates, and that is why it dislikes *"qualitative things"*. Quality is subjective and cannot be proved from that "objective reality" perspective. Physical things are measurable. Easy. A Newtonian mentality. The bottom line is that, if we are not really interested in knowing what it is like to be a bat, we should think of reality in terms of our "collective" reality. The reality all human beings share. Reality is totally different for the different species. That does not need proving, as sensorial capabilities vary from one species to the other.

The current trend in the scientific world, which is clearly biased towards physicalism, tends to minimise the role culture and cognition should play in the study of consciousness. But, the phenomenon of consciousness is not just sentience. It is possible to deceive yourself into believing that by researching sentience you will somehow arrive at the whole of consciousness. That is an impossibility.

In his book *Being you*, Anil Seth reiterates the current scientific position several times, i.e.:

> *"The beast machine developed in this book makes the case that consciousness is more closely connected with being <u>alive</u> than with being <u>intelligent</u>. Naturally, this applies as much to other animals as it does to us humans. On this view, consciousness may be more widespread than it would seem, were we to take intelligence as the primary criterion. But it does not mean that wherever there is life there is also consciousness."*

The approach is short-sighted and it ends up not explaining anything. Not only that, in its extreme version, it adopts old nonsensical ideas like panpsychism. It's not only that the position does not explain anything. We can see the result in the many books and articles produced by neuroscientists, philosophers, and researchers, which finally confess that they have not even begun to understand the "mystery" of consciousness. At least by including cognition and culture, research can be directed in a clear direction. There is no doubt in my mind that—contrary to what

Seth opines—a new definition of consciousness is urgently needed to clarify matters.

Seth, and many other neuroscientists, do not want a definition. They propose: *"Any kind of subjective experience"*. That is not a comprehensive definition of consciousness. It excludes what is unique to human consciousness. But he explains further his clearly untenable position:

> *"When a complex phenomenon is incompletely understood, prematurely precise definitions can be constraining and even misleading"*.

The problem I see with that is that, if you begin by plainly admitting you ignore the object of your study, the phenomenon is not only misunderstood, but you don't even know what you are trying to understand. Of course, a more comprehensive definition does not need to be inflexible forever. Many amendments normally come up as you progress with your research. But you need to start somewhere, and that place has to be clear.

The way the issue has been raised means that all of the scientific world takes one thing for granted: human consciousness originates in the individual brain. The process is purely physical. That is an unacceptable way of explaining how communication occurs. It would presuppose a monolinguistic environment. Every individual would be able to communicate with the rest of humanity. Or, the coincidence would be that the individual would choose to communicate with another individual who spoke the same language.

Patrick House (*Nineteen Ways of Looking at Consciousness*, 2022) says exactly that:

> *"There will never be a scientific model that explains all of biology or all of the brain or the way in which neurons generate consciousness or are themselves conscious".*

"Neurons generate consciousness. Neurons are themselves conscious". That is an amazing statement, totally unfalsifiable. And he, like many other neuroscientists, take it as a given.

I might be oversimplifying the issue, but, from my perspective, any attempt at explaining human consciousness without culture is doomed to failure. Historically, there have been explanations of social structure, like the one provided by John Searle, but there appears to be no current link with anything happening today in the field of consciousness.

≈

IN THIS VERY AMBIGUOUS FIELD, I believe no goal can be achieved without consensus of some sort. What do consciousness researchers want to study and for what purpose? What do they want to find out? If they are interested in animal behaviour, by all means, that study should be circumscribed to the material properties and processes of the brain. The brains of most mammals and those of other species have evolved uninterruptedly from a biological perspective. Human brains and human minds are markedly and qualitatively different; therefore, if those same researchers are interested in _human_ consciousness, in phenomena that include long-term memory, imagination, creativity and sophisticated communication processes—which, to some extent, are lacking in other species—the only way that aim can be achieved is by including a cultural component.

That cultural component is the most important element in the human mind. It is the one that has made humans incredibly successful; it is the one that allows us to co-operate, to imagine, to remember, to learn, to create.

This book, at least, is about human consciousness. It's about _psyche_.

∽

FOR THOUSANDS OF YEARS, human beings have been fascinated by the idea of human consciousness, or, as it was originally called, the _"soul"_. Many recent hypotheses were forwarded about its nature. Some were fairly sound. Why is it that it is still an unresolved problem? Well, for starters, consciousness has never been clearly defined. Its complexity lies in the fact that it appears to reside within our mind, but it has no border—it can go outside of our skin—because we can see and hear, but also communicate very complex ideas to other human beings. There is a dual nature to the cognitive part itself, in that it is individual and, at the same time, collective. The fact that you are reading this text now is evidence of it. You are interpreting ideas that I am proposing to you in writing.

Sometimes, we can even communicate or give simple commands to individuals of other species.

The observer that lives—or appears to live—within our head is difficult to define, or to study from a non-humanistic perspective. That observer (what used to be called the *"homunculus"*), is sentient and, at the same time, it is a thinking individual that can also communicate with other individuals. I repeat, it is subjective and cultural at the same time. It is part of a collective.

We can perceive things that happen inside of our bodies, i.e., interoception; and things that happen outside, that is, exteroception (percepts of stimuli that originate outside of our body). We are self-aware and, at the same time, we perceive outside objects and events.

We were saying that we can also think and communicate: if I feel sick, and if you speak my language, I can explain my symptoms to you. I can also explain to you that I see a red butterfly in the garden. Under normal circumstances (and with possible variations in the way you yourself feel and perceive) you will understand me. That clearly involves culture and language. My illness and my perception have internal and external physical components, but the exchange of information about them is a culturally and linguistically superimposed explanation of feelings, or sensations, of objects or events.

We have an identity, we have memory, we have imagination, we have free will. If we were only collections of atoms and particles that obey the laws of physics, we would not have free will. Instead, on the basis of those abilities, we can act.

∾

Thomas Nagel, in his book *"Mind & Cosmos -Why the Materialist Neo-Darwinian Conception of Nature is Almost Certainly False"*, argues that:

> *"If evolutionary biology is a physical theory—as it is generally taken to be—then it cannot account for the appearance of consciousness and of other phenomena that are not physically reducible. So if mind is a product of biological evolution—if organ-*

isms with mental life are not miraculous anomalies but an integral part of nature—then biology cannot be a purely physical science. The possibility opens up a pervasive conception of the natural order very different from materialism—one that makes mind central, rather than a side effect of physical law."

～

WE ARE SENTIENT BEINGS. Sentience is our genetic ability to feel, to perceive, pretty much the way other animals do. It's part and parcel of being alive. In fact, it's a basic requirement for staying alive. Let's call it *nephesh*, which is a biblical term for "breath of life". Later, I will explain why (by the way, there is nothing religious about the adoption of the term).

I repeat, my perspective is that there are two levels—two discrete layers—of human consciousness that have a totally different origin and nature, but they can interact. I also believe this is clearly explainable and needs no further verification as it is self-evident. The layers appeared at different times in response to different needs of our species.

According to Aristotle, we are also social animals; "*zoon politikon*" he called us. Basically, we need society to survive. Infants need the collective. Sometimes, individuals cease to function in society; we take that as a symptom of a mental pathology. Our human consciousness, our thought processes, developed, and continue to develop, culturally, in every human individual: it is a higher form of consciousness that goes beyond biological consciousness, let's call it *"psyche"* (that is a *Koine* Greek term that originally meant something like *"soul"*). Individual children are born without it. They acquire it as they grow up. Again, later we'll see why and how.

How did we develop *psyche?* Well, we need to communicate with other members of our species, especially and mainly within our culture. Because of that, we developed language, which was plausibly the beginning of becoming human (*Homo Sapiens*). We are also a weak and altricial species, that is, young individuals require many years of being looked after by their parents and other members of society before they are ready to fend for themselves. That way, language and culture are passed on from generation to generation. The process ensures that a great deal of information is retained by the collective. In other species, knowledge is completely lost with every individual that dies. All that remains in those species is laboriously and slowly acquired instinct.

As human groups grew from families, to clans, to tribes—largely as the result of food production and sharing—, and ultimately to nations, communication necessarily expanded in complexity and sophistication. Our individual *psyche* grew, accordingly, in exponential fashion.

But we remain a highly social species after childhood. Even after individuals have grown into adulthood, there is a direct correlation between brain activity and social interactions. Not only do good social connections provide feelings of wellbeing, but being understood by others is very important in terms of greater life satisfaction. Even bad social connections sometimes induce learning. In order to reach a fully developed *psyche*, humans need the approval of the collective.

Feedback to/from society—sometimes called "global consciousness"—and the individual remains a powerful force in terms of the development of cognition in the individual, and in turn, this results in benefit to the community. Social and cultural interaction, then, is fundamental in the development of the physiology of the brain: brain activity benefits from society and vice versa; and it is only indirectly related to evolutionary or physical biology. Although these interactions can be measured to some extent, they are not physical, they are purely cultural.

CHAPTER 2

❧

THE WAY THE STORY HAS DEVELOPED

*"One glance at a book and you hear
the voice of another person,
perhaps someone dead for 1,000 years.
To read is to voyage through time."*
- Carl Sagan

*I*n the beginning—even without God—the planet was in "biblical" mode. Land masses and seas went their own way and, once they were apart, everything was quiet; everything sat still on planet Earth. There was neither observer nor observed. Noise could have occurred, but even that process wasn't complete as there was no listener.

Seven hundred million years ago, or so they say, somehow, something changed. There was an instant when there was molecular change: life. A tiny speck, life began reproducing, and did so aggressively. For aeons, life grew everywhere, but unwitnessed. Plants and trees grew tall, grasses populated plains, living creatures moved and reproduced. There were fights, predators and prey. Life was raucous and ubiquitous, omnipresent. Still unwitnessed. There was change, but no consciousness to witness what was happening.

Between nine point five and six million years ago, hominins diverged from the closest primates, the chimpanzees. Much, much later, some tens of thousands of years ago, life was becoming more and more complex. Those ape species that had split from the chimpanzees—the hominins—lived in clans, for protection against predators and other dangers. Their offspring had to be cared for over long periods before they could look after themselves. These hominins, these apes, moved and lived individually, but needed to have their group to survive. They were empathetic towards their group and its needs. They looked after each other, and groomed and helped each other.

One of the first hominin ancestors we are aware of was Lucy—an early female australopithecine who lived in what is now Ethiopia some 3.2 million years ago (to put her discovery into context, as we discussed, humans and chimpanzees had a common ancestor between 9 and 6.5 million years ago). Lucy was fairly short, bipedal and upright. Her skull and brain were small (275ml) compared to ours (1400ml).

Some argued that, although bipedal, her species was still partially tree-dwelling. Lucy appears to have been closer to other primates than a current human. But one interesting discovery about her species is that their brain took longer than those of other apes to reach adult size. Their infants were dependent on care-givers for longer periods before reaching maturity. We appear to have inherited that trait. I would venture that our weak origins, combined with our altricial nature, therefore, were instrumental in the development of human consciousness.

∼

THEN, something even more amazing than life happened. An individual in one of those groups solved a problem. The group learned to copy the solution. Eventually, there were tools; there was thought, and with it came a basic form of communication among peers: the germ of language.

An animal had become someone, a human being. Slowly, individuals in human groups took on identities. They noticed details in their surroundings. They learned from each other, and remembered what they had learnt. They passed their knowledge to their offspring. Information was shared. Then, the corpus of information shared by humans grew exponentially.

Research conducted by archaeologist Ludovic Slimak during the past thirty years tends to confirm that

Homo Sapiens passed technical information from one generation to the other over tens of thousands of years. Slimak and his team found that *Sapiens'* tools or flints were more efficient and more homogeneous than those crafted by *Neanderthals*.

After analysing many *Neanderthal* tools, Slimak could confirm that they were all completely different, that each tool was a specific creation. The information *Neanderthal* hunters possessed—then—like those of bears, wolves and other mammals, appeared not to have been transmitted. They were literally "reinventing the wheel" every generation. Humans, on the other hand, copied the more successful solutions to problems.

Human clans may have interbred with *Neanderthal* clans, but what becomes clear from Slimak and his team's discoveries is that *Neanderthals'* humanity was not like *Sapiens'* humanity. They might not have been transmitting information and developing some form of *psyche*, like *Homo Sapiens*.

Of course, at the beginning, humans were hunter-gatherers. Archaeological studies suggest that— possibly with the advent of agriculture—human groups grew, and the knowledge of individuals could be preserved even better. Information spread among tribes and was passed orally through generations.

Language grew more complex; with it, explanations became recursive, and conscious reflection appeared among these primates. An animal species had begun to witness life. *Homo Sapiens* had acquired consciousness.

∽

WE HAVE scant information about the daily life of our ancestors before oral history and literacy. New paleoanthropological research keeps on uncovering traces of their diet, habits, sex life, interbreeding with other species—like *Neanderthals* and *Denisovans*—, genetics, arts and crafts, and wanderings around the planet.

As the social species that we are, we know that our ancestors lived in small groups, probably extended family groups at first; later in clans—or groups of interrelated families—; and later, when trust and food sharing permitted it, in tribes: even larger groups that shared a language and developed a common culture. To be able to achieve communal living of that nature, humans had to establish some type of government, probably a leader or king, and ethical and moral rules that determined what was good or bad for the group.

There is total certainty that our first human ancestors were hunter-gatherers that led very short, hard lives. In his epic poem *"De rerum natura"* (*"The way things are"*), Lucretius, the first-century CE Roman poet, imagines how the first humans lived:

"Of sun withdrawn forever. But their care

Was rather that the clans of savage beasts
Would often make their sleep-time horrible
For those poor wretches; and, from home y-driven,
They'd flee their rocky shelters at approach
Of boar, the spumy-lipped, or lion strong,
And in the midnight yield with terror up
To those fierce guests their beds of out-spread leaves."

Lucretius did not refer to our human ancestors as apes, but he did make it quite clear that their lives were not much better than those of other animals.

How were their lives and their social environment? Other large ape species—like chimpanzees and gorillas—can teach us a lot about the lives of our ancestors before language.

Jane Goodall travelled to Nairobi, Kenya, in 1957. There she started working as an assistant to Louis Leakey, the famous paleoanthropologist who demonstrated that humans had originated in Africa. Goodall commenced working among the Gombe chimpanzees some sixty odd years ago. She studied their social and individual lives. Thanks to the many things she discovered, we now know that chimpanzee offspring are dependent for approximately nine years, which is approximately half what human childhood lasts; human children require care for longer. Females generally look after their babies and gather food from trees. Chimpanzee clans are loosely governed by alpha males, who make decisions for the group. They control the territory, travelling habits and even the food the group eats. Chimps are very sociable—like us—and each clan develops separate cultures. Even though individuals appear to copy the use of tools, like sticks to eat termites from their nests, their tools are not crafted and there is no need to pass detailed information from one generation to the other. They have limited creativity and, of course, a limited form of communication.

Through the research conducted by Goodall and her team—which continues to this day—we can make many educated guesses regarding the lives of primitive humans. Of course, there are differences with the lives of current primates. Our ancestors migrated in groups and, when they were fully bipedal, did not need trees for protection or as a habitat. They lived in clans and were mostly cave-dwellers until they developed the technology to construct their own huts.

One fundamental difference between our ancestors and other primates living in groups was pointed out by researchers from the University of Minnesota, led by Professor Michael Wilson: the origin of food sharing among early humans. The presence of pair bonds had a very important impact on the food production and sharing. Males in pair bonds protected females from other males and that, in turn, incentivised food gathering efforts by females. The females could concentrate on gathering high-quality food, which they shared with their mates. The research group suggested that the presence of pair bonds among humans was crucial in the evolution of food sharing in human society. That enabled our human ancestors to move to different habitats and helped them with the survival and reproduction of offspring.

~

THE EMOTIONS, feelings, development of the individual minds of our ancestors, again, can be guessed through studies like those conducted by Allen and Beatrix Gardner, two American anthropologists who, in 1967, taught a female chimp—Washoe— how to speak Ameslan, the American sign language. Washoe was the first non-human individual who could communicate using a language. She taught sign language to her adopted child and, later on, as she was introduced to other chimpanzees, she taught them Ameslan as well. Washoe demonstrated without doubt that she had a high degree of self-awareness and emotion. Many other studies involving communication with other primates, like gorillas, were conducted afterwards. The results obtained through those studies have been encouraging.

David Graeber and David Wengrow (*The Dawn of Everything*) propose that prehistoric societies had clear hierarchies and that those hierarchies varied according to the different groups.

The appearance of recursive language and a germ of consciousness seems to have occurred some tens, maybe hundreds, of thousands of years ago in Africa and later in Europe, at different stages for different groups.

The myth of the Tower of Babel— which served as an original explanation for the diversity of languages— was very apt for its time, but the real origin of language appears not to have been monogenetic. It must have occurred simultaneously in different parts of the world. As for consciousness, well, we have seen limited creativity in other species that use tools—like crows and chimpanzees—; and we have seen that human language could be transmitted from one or two speakers and grow exponentially after that.

In his "Romulus and Remus" hypothesis, Andrey Vyshedskiy proposed that, some seventy thousand years ago, a genetic mutation in two children caused the slow development of their prefrontal cortex and originated recursive language.

What appears clear nowadays is that the introduction of food production and sharing—agriculture and animal husbandry—was a pivotal moment in the development of culture. Like other animals living in small clans, humans transmitted and preserved certain amount of knowledge, maybe even before then. When agriculture allowed long-term human settlement, and trust and interbreeding fostered contact among clans, sophisticated knowledge began to be preserved for future generations and could be transmitted to other tribes. That period marks the beginning of long-term memory.

The recent excavation of a village in Israel challenges the traditionally agreed date for the beginning of human settlement. Apparently, a culture called the Natufians cultivated plants, built stone houses and had cemeteries some twelve thousand years ago. The excavation of the village discovered advanced tools and artifacts, as well as face carvings and decorative beads. Regardless of the date when that happened, food production was a decisive factor in human settlement with its consequences on the exponential growth of human consciousness.

∾

WE KNOW that many changes happened to humanity before we were born. Some were recorded, others weren't. Some changes were lost to history. Some were remembered orally for generations, as we discussed in the previous chapter. But, —and this is very important—in order to explain reality with all the changes that happened before us; in order to explain our long-term memory, we *invented* time.

When we say "time immemorial" we are talking about a time that existed before our memory. We have no evidence that it existed. In fact, there was no such time. All the records we have are archaeological. Before human consciousness, our ancestors were just like all other animals. There was change, individuals lived, died and were born, but without individual or group identity, nobody kept track of when those events happened; nobody knew who did what. Time was not measured because nobody knew anything except change. Information and communication were scarce.

The beginning of time is simultaneous with the beginning of long-term memory. All there was before memory was change. When we had not developed culture, when we were like all other animals, in effect, we had no time.

According to recent studies, the first—plausibly mnemonic—notations about seasons and parturition of prey were recorded in caves throughout Europe some 40,000 years ago. Before then, nothing. Our primate ancestors before the Upper Palaeolithic had no notion of time because their consciousness—their *psyche*—was in an early process of development. I would venture that, with the advent of food production, about 10 to 12,000 years ago, there was an explosion of language, consciousness and culture. And, of course, of time. No other species appears to have time or understand the concept of time.

Time—according to Aristotle—is the measure of change. We also need time to live in society, with other human beings. We'll discuss that in a chapter exclusively about Time.

~

WHAT FOLLOWS IS a brief discussion of a connection between religion, philosophy and science in the West.

You are probably thinking: Why do we digress here? How is religion remotely related to the study of consciousness? How will this lead us to consciousness? Bear with me. There is a non-religious, historical link between oral history and religion and that, in turn, will take us to the way Western science and philosophy have studied consciousness for a long time. We must say—though—that the way Western science has studied consciousness, as presently conceived, has not been very successful thus far.

Let us try to look into the reason for the current situation in terms of our understanding of *psyche*.

Of course, these are things that science finds hard to explain. Those phenomena call for a different type of understanding. Religion, or theism, finds ways to explain the existence of nature in ways that are not provable or falsifiable. We are not discussing those here.

Historically, religion—or an element of it—was an attempt to explain something our ancestors didn't know or understand. Let's put it differently: theism is a primitive form of science, a quest for knowledge. Human beings are naturally curious. For millennia, human beings asked questions about their origins and how they came to be. When you do not have a logical, verifiable answer, the obvious solution is to mythicise. Many cultures did, and still do, exactly that; many civilisations came up with answers, some far-fetched, some, credible for their time. The Egyptians were one such civilisation. The Hebrews were another such civilisation. Among other ideas, they came up with monotheism. The latter interests us because we have a comprehensive written record of the way they saw their origin and the origin of humanity, and because Christianity—the religion that became predominant in the West for many centuries at the end of the Roman Empire and after its fall—derives from Judaism. Of course, there are other, ancient links between culture and religion.

In Greek mythology, for instance, the nine muses were the daughters of Zeus and Mnemosyne, the goddess of memory. Greek tradition, then, tried to explain the existence of art and culture by recognising the importance of memory. Mnemosyne was one of the Titans, who symbolised a past that could not be explained (prehistory?). The Gods triumphed over the Titans and Mnemosyne ended up conceiving the muses, who began their lives as nymphs and then grew into adulthood and into their own fields. There is a lot of poetry and eloquence, music, and theatre;

but the arts also included history and creativity, geometry, agriculture and science. There was an attempt by Greek mythology, then, to assign these symbols to activities that were cultural and human, and were separate from more basic, physical, activities, like sports. For the ancient Greeks, then, symbols were divine, and only humans could understand symbols.

～

BUT LET us go back to the Hebrews. The ancient Hebrews, like many other peoples, had oral traditions that had been transmitted for centuries through song, at home, and around campfires. The traditions, passed from one generation to the other are now known to us. We know about King David's triumph over Goliath, and about the wisdom of King Solomon, and the Judges, and Samson. We know about Sodom and Gomorrah, and how the Walls of Jericho fell down on the seventh day, when Joshua's warriors cheered and their shofars sounded a great blow. How do we know all of these legends, or traditions, if they were transmitted orally?

Religion—we said—tried to explain something unknown. One of the intentions of King Hezekiah when he ordered that the Bible be written was to produce a transcription of the oral history of the nation.

Like many peoples of their time, the Hebrews were an illiterate nation. The first texts on stelae and inscriptions on marble and rock are from around the 8th century BCE. That was the beginning of writing in Judea. Of course, early writing had a divine nature: only a few could read and write; the nobility, some priests and court scribes could do it, the rest of the population were illiterate. Coincidentally, in the 8th century BCE, King Hezekiah was facing a crisis. The Davidic kingdom of Judah had been divided into two kingdoms, Judea and Israel. Assyria had occupied the northern kingdom—Israel—, many had been killed, many deported, and many had escaped, especially from Samaria, the capital, to Jerusalem, the capital of Judea, and surrounding areas. There was overpopulation, there was unemployment and lawlessness in Judea. To impose the rule of law again in his kingdom, Hezekiah needed more authority from God, a code of law, and a history. The idea was to compile all traditions (the oral history they already had). The scribes had to explain how God had created humanity, how the Jews were the nation God had chosen from the beginning, how God had given them ethical and legal principles; they had to assemble a history from all their traditions, and write them down (Scripture). The result was the Bible. It was the first attempt at a comprehensive history of a nation, maybe not the way we see history nowadays, but history nevertheless.

To be credible, the Bible had to explain everything from the very beginning, not only from the origins of the Hebrew nation, but from the origins of humanity. And King Hezekiah's scribes did exactly that. They provided not one, but three accounts of how humanity started. In the first one, God creates everything, from light, day and night, the Sun and the Moon, the waters, etc. And then, he creates humanity.

Three important things happened at this point:

1) God gave human beings "dominion over everything else";

> "26 ¶ And God said *Let vs make man in our Image, after Our likenesse: and let them haue dominion ouer the fish of the sea, and ouer the foule of the aire, and ouer the cattell, and ouer all the earth, and ouer every creeping thing that creepeth vpon the earth.
> 27 ¶ So God created man in his owne Image, in the Image of God created hee him; *male and female created hee them" - (First Book of Moses, King James Version).

2) God created humans and other living beings by giving them *nephesh* "the breath of life". These were sentient beings. As we have seen, *nephesh* is an ancient Hebrew term for "sentience". So, God gave humans sentience and also made them better than other animals and separate from the rest of nature;

> *"7 And the LORD God formed man †* of the dust*
> *of the ground, & breathed into his nostrils the*
> *breath of life; and *man became a liuing soule." -*
> *(First Book of Moses, King James Version).*

3) God allowed humans the possibility of reasoning, but warned them that, the moment they could tell good from evil, that would make them also aware of their own finitude;

> *"15 And the LORD God tooke | | the man, and put*
> *him into the garden of Eden, to dresse it, and to*
> *keepe it.*
> *16 And the LORD God commanded the man,*
> *saying, Of euery tree of the garden thou mayest †*
> *freely eate.*
> *17 But of the tree of the knowledge of good and*
> *euill, thou shall not eate of it: for in the day that*
> *thou eatest thereof, thou shalt † surely die."*

To summarise, the Tanakh, the Hebrew Bible, established three important principles: 1) Humanity is separate from the rest of creation (objective reality); 2) God creates life (nephesh); 3) God gives human beings the possibility of becoming conscious and ethical (knowledge of good and evil); this is what happens according to the Tanakh, the Hebrew Bible.

But there is much more to the myth (Pintos-López, R - *The Myth of Adam and Eve and the Endurance of Chris-*

tianity in the West, 2021). Let me give you one tiny example of the sophistication of the authors of the Book of Genesis and the complexity of the Myth of the Garden of Eden:

> *"16. Unto the woman he said, I will greatly multiply thy sorowe and thy conception. In sorow thou shalt bring forth children...".*

Why did God tell Eve that at that point?

Adam and Eve were becoming human. The neocortex in the human brain is thirty times larger than that of the closest mammal. Women appear to be the only mammalian females that give birth with pain. Were it not for the fact that the brain has to fit within the skull, and the fact that the cortex has convolutions, with *sulci* and *gyri* (folds and creases), the human brain would be much, much, larger. In any case, it is very close to the size the birth canal can allow it to go through. Caesarean incision births are on the increase, and there is a reason for that (the size of Lucy's brain, if we remember, was about one third the size of the brain of a current human being).

∽

TEN CENTURIES LATER, Christianity calls the Tanakh "Old Testament" and adds a "New Testament". The New Testament originates in the life and work of Saul of Tarsus, a Jew from the Greek diaspora. We are not going to enter into how Saul (aka St Paul) became the leader of an unknown peripheral cult of Judaism. Let's say that Paul was literate, multilingual (he spoke Greek, Latin, Aramaic, and possibly Hebrew) and familiar with Plato and other Greek thinkers. He did not write the New Testament, but his ideas were the foundation of Christianity.

The New Testament establishes a difference: *nephesh* is the "breath of life", something all animals have. In this new addition to the original book, instead of "knowledge of good and evil", humans have something that is represented by a Greek word: *psyche* (a "soul", as opposed to animals).

Christianity clearly established that sapience was a separate entity and added it to the equation. This tradition had a fundamental importance in Western approaches to nature, the world and the cosmos, and profoundly influenced the development of science.

The new religion is based on the immortality of *psyche.* All human individuals possess an individual immortal soul. One element of the Holy Trinity is totally divine (God Father); another element is partly human (Jesus is a man); and the Third element is the divine side of humanity, the Holy Spirit (Human Consciousness). Thus, Christianity provides a religious explanation for collective consciousness.

What that means is that human beings are semi-gods. They have an immortal *psyche* (what we now call consciousness). They are separate from other animals and the rest of nature. Human beings are creators, very close to gods. Actually, Jesus Christ is a man. Humanity can study the universe from the outside. They own a reality that is "objective" because they are separate from it. As mentioned above, there is also a collective human consciousness that needs to be added to the equation: hence, the Holy Spirit.

Christianity also provides an explanation for free will: unlike other species, we can make good and bad decisions and are accountable to God. We can act in a good or a bad way and have to face the consequences.

Under threat of death, the West faces a few centuries of unchallenged religion. After that, Western civilisation gives birth to science out of a mixture of philosophy, alchemy, cabbalah and superstition. Isaac Newton—the last of the magicians, an alchemist—sets the foundations of Western science. He discovers gravity and changes the rules of research.

In the nineteenth century, Charles Darwin writes *On the Origin of Species* which establishes, beyond any doubt, that all animals, humanity included, have evolved from lower, less sophisticated species. Physicalism and evolution appear to explain everything. The dualism that Descartes advocated three centuries before does not explain the origin of the human mind any longer.

The genealogy and philosophical leanings of Western science, then, are quite clear, up to Einstein. The irruption of quantum mechanics in the early twentieth century is about to change everything.

Well, we all know Copernicus established that our planet revolved around the Sun, and Galileo continued with his heliocentric ideas a century later. However, our Western minds still tell us that we are the most important thing in the universe. The West is not as Christian as it once was, but Christian philosophy is still a large influence on science, ethics, and Western society in general. We behave as if human beings were still separate from all other animals and had God-given dominion over everything else.

As we have seen, Charles Darwin demonstrated that our species, like all other species, had evolved from lower forms of life. One thing he did not explain was human society. In any case, the Christian view that the human species is in charge of the planet clearly persists in Western culture.

There are other interesting notions about oral history and conclusions we may reach from the way it becomes Scripture. Oral history is a first attempt by human beings at remembering events that happened before their lifetimes. After a few generations, events are distorted, exaggerated, aggrandised. Actual historical episodes become legends. Heroes become super heroes; battles are won by a miracle of God; events are epic: the Jews take forty years to cross the desert; God talks to Moses; Moses' face is so radiant that nobody can look at him directly.

Forty years is a way of measuring time; but, to build the ark, Noah also receives precise instructions with measurements: 300 cubits long, 50 cubits wide, and 30 cubits high (that is more than one hundred metres in length, for instance). It had to be a large ship. It gives a clear idea of a large ship.

Although, in general, its descriptions are fairly imprecise, the Tanakh includes measurements of time and space. Another beginning of objective reality.

The Ark of the Covenant—which was of fundamental importance to Judaism—also had to have very specific measurements and had to be constructed in a precise manner: it had to be covered in gold, etc.

So, something that is also interesting about that oral history—which we now know thanks to the fact that it became scripture—is that it is generally fuzzy. Like the memory of a child, like the consciousness of a child. It starts with a guess about something that is totally unknown: God creates everything from the beginning. Then, there are periods when countless unknown—or barely known—generations pass, like those between Noah and Abraham.

According to the Bible, humanity begins with Adam and Eve, and with the fact that they become conscious. They can tell good from evil. It follows that, before the episode with the snake, they were sentient, but not fully conscious, thus, not quite human. My reading of the story is that King Hezekiah's scribes were proto-Darwinian. Some will not agree.

∼

HISTORY TEACHES us about things that people did, changes that happened long ago, even well before living memory. We're used to it, but how does it happen?

Other animals have no history. Other animals have not developed long-term memory either. They have short-term episodic memory and live constantly in the present, in the "here and now". They don't know they are going to die. When you think about it, they are immortal. They only understand immediate change and have limited learning capacity.

Changes happen to bears and foxes and, as we said, they learn from change. But when a good hunter bear or fox dies, the next generation of bears or foxes will have to learn all over again. We—on the other hand— have long-term memories and can pass information from one generation to the other, many times over. We learn. We create. Our knowledge grows exponentially. We know about Roman emperors, Visigoths, the French Revolution, among other events and peoples. As a culture, we remember important changes that happened many generations ago.

So, what is history? History is a cultural invention partly derived from literacy.

Thanks to history we know, for instance, that Nicolas Malebranche was interested in consciousness—just like you are. Only he was a Catholic priest and lived in seventeenth century France.

In those days nobody talked about "consciousness" as such. People had a spirit, or a soul, and God was pretty much involved in the way that soul operated. In the 1670s, Malebranche wrote a book, *De la recherche de la vérité (On the Search for Truth)*. In the book, Malebranche studied—and defended—Descartes ideas on metacognition, that is, why and how we think about thinking. In other words, ideas on how we think and why we do it.

René Descartes had recently published his famous *Traité de l'Homme (Treatise on Man)* in which he explained his ideas on the duality of the mind (the fact that body and mind needed to interact, but one half was matter, and the other half, non-matter). Descartes decided that they operated separately. His ideas are known as Cartesian Dualism.

We have to remember that, in those days, saying that there was no soul, that consciousness had originated in matter, would have been an open attack on the Church. These days, it is exactly the opposite: saying that consciousness has originated in culture appears to be an open attack on neuroscience. Now, they won't burn you at the stake, but they will ridicule you out of existence. Some things are taken for granted and there is no room for doubt. Dogma, apparently, continues to reject new ideas.

Descartes reasoned that being certain of existing as a cognitive individual, separate from his physical body, helped him understand that the mind was a distinct soul. His approach was totally different from those of other European philosophers of the day. He invited people to look at reality from a first-person, fresh perspective. Moreover, he encouraged doubt as a fundamental way of advancing science. Everything had to be questioned. By questioning, he verified even his existence. *"Dubito, ergo cogito. Cogito, ergo sum"* (*"I doubt, therefore I think. I think, therefore I am"*).

∼

MALEBRANCHE AGREED to some extent with the idea of dualism, but thought that maybe human beings could function, in spite of the dichotomy, because God was the intermediary between soul and body. He was probably referring to the Holy Spirit. We now know the Holy Spirit is the Catholic equivalent of cultural information: what we nowadays can find in Google.

Baruch Spinoza was a seventeenth century Dutch Jewish philosopher of Portuguese descent. He believed that there was only one substance. Mind and matter, according to him, were one and the same. Spinoza's idea of consciousness was a mixture of monism and panpsychism: the mind is part of nature (or is nature), and at the same time, it is the universe and God.

Berkeley, an eighteenth-century philosopher, also a priest, but this time from England, came up with another idea: reality only exists within consciousness. Things do not exist without being observed. Many scientists appear to be returning to that notion. It's called Subjective Idealism.

Karl Popper, the famous philosopher of science who devised the concept of "falsifiability", proposed an ontology that divided reality into three "worlds": the first one included everything in the animal and vegetable kingdoms, and everything in the cosmos; the second one had things like feelings like pain and pleasure, but also thoughts, ideas, fears and hopes (nothing objective); and the third one included all products of the mind: objective contents of thought, like science, poetry, and art.

What Popper had in mind appears to have been a taxonomy that involved objective reality, subjective human consciousness, and a mixture of both realms (human creation and inquiry). His hypothesis includes a division of sorts between biology and culture. But he could not quite grasp the cultural origin of human consciousness, which would have provided a distinct division and clarified the subject.

As we have seen, an Australian philosopher and cognitive scientist, David Chalmers, formulated a question he called *"the hard problem of consciousness"*: *"why do we have awareness of sensory information?"*, i.e., a question that, in my view, should result in a dualist answer. Our bodies are matter, but we feel and think. How does that happen? How do we have a conscious subjective experience? But the way the question is formulated, it is not conducive to a dualist answer.

Chalmers' question was formulated so as to include sentience in the brain. That way, the answer has to be physicalist. *"Why do my neurons and the particles in my brain produce sentience?"*. The question only addresses the existence of consciousness in the brain. The answer it would elicit would be physicalist, as it does not include the existence of cognition in the individual or in the collective. The question is inappropriate. And it has no answer because neurons do not generate consciousness. It is the other way around. Culture and behaviour generate genetic change, which produces the corresponding DNA, neurons, synapses, particles, etc.

Through the years there were many philosophical and scientific attempts to explain consciousness. Questions such as Chalmers' remain unanswered. Our conscious subjective perspective remains unanswered.

In philosophy and science, things appear to go in and out of fashion. There is a pendulum, whereby one paradigm becomes *the* solution until somebody falsifies that solution and comes up with a modified version of it, or with the exact opposite.

In the decade of the 1980s, philosophers and scientists coincided in that they were interested in brain and mind research. It was the beginning of neuroscience. The good thing about it was that during those days there were many multi-disciplinary attempts. There are still meetings, but real communication among the disciplines appears to be dying down.

Absurd as it sounds, some neuroscientists keep on trying to find a place in the brain where consciousness is generated and stored. I believe by now we all know that attention, thought, perception and episodic memory occur, to some extent, in the neocortex, and that the thalamus has many functions that include sleep, wakefulness, memory and learning.

Psyche appears to be integrated, though, and what those neuroscientists are doing reminds us of 19th century phrenology, when people divided the brain and drew maps on those funny illustrations. Human consciousness involves a degree of integration within the brain, but it is also discrete and cannot be explained solely in physical terms. The moment you understand that *psyche* has a cultural origin, the whole physical search becomes clearly inane.

Having said that, an interesting recent discovery is that memory might be located in the membranes of neurons. Synapses transmit information from one cell to the other. Between synapses there is what is called the synaptic cleft. Memory and learning are stored there. One neuron sends information and another neuron receives it. The change that occurs is called synaptic plasticity. That can be short or long-term. Chemical processes that take place between synapses and create short-term plasticity may potentiate it into long-term plasticity: learning and memory. Our species must have developed long-term synaptic plasticity through social interaction.

CHAPTER 3

✲

"OBJECTIVE REALITY" AND WESTERN SCIENCE

*I*n the previous chapter we saw how the Western concept known as "objective reality" was introduced through religious and philosophical ideas. It grew out of the notion that human individuals were separate from everyday reality. They could observe reality from the outside. Some would say that objective reality came after Plato. Yes, that is a possibility, you could say that. I would say that it came after Christianity. I would argue that Christianity—actually a child of Judaism—and science are opposite poles of the same paradigm.

Some would allude to a shift in Western thinking that occurred around the time Jaspers called *"the Axial Age"* (about 500 BCE). In those days, men like Anaximander and Thales moved away from religious or mythological thinking into more rational thinking. The fact remains that the individual immortal soul that Saul of Tarsus imagined for theological purposes was the same as the individual immortal *psyche* Plato had thought of. Western science has an undeniable religious genealogy. Not that there is anything wrong with that. We all know that Christianity also owes a lot to Greek philosophy. The only problem that remains is that the objectivity of reality is questionable whether it was proposed by Anaximander, Plato or Saul of Tarsus.

In any case, the result is that—as things stand today—Western science can observe, analyse, and study reality, based on the assumption that there is an objective reality against which any idea can be tested. It is a big assumption; as we know: our senses are quite limited compared to those of other sentient beings.

I am sure, the way we intuit the "objective" physical world that surrounds us is somewhat similar to the intuition of a dog, that has no notion of time, and whose sense of smell is two hundred times ours. The dog believes—it would make sense—that its world is the way reality is. When it is among flowers, a humming bird, which perceives a million colours, probably intuits that flowers are part of an incredibly sophisticated chromatic world. We cannot even imagine that reality. But, without going beyond our species, I can ask myself: would Mozart have understood my extremely limited sense of rhythm, melody or tempo—for instance—from the perspective of his incredible musical talent? Yes, those things are measurable, but to what extent? No doubt, perceptions and sensibilities vary among species, but also among cultures and individuals. Can reality be objective? Well, according to quantum mechanics the question is a valid one.

I am sure that for that sort of question there is a middle-of-the-road, Kantian explanation aimed at scientific objectivity, whereby our mental representation is the one that makes the object possible. According to that view, our mind would actively originate the experience rather than being a passive recipient. In any case, constructing a world through our mental frameworks, perceptions and categories would not make that reality really "objective".

That kind of explanation would assume that our view of the world is based on our biology, and that human beings have had those frameworks built into our minds through hundreds of thousands of years of evolution, which would make the frameworks and the perceptions part of the real world. Again, if we assumed that, we would also need to discuss Berkeley's idealism, Hume's sceptic ideas, and all kinds of philosophical questions, and that would only confuse the issue.

In fact—as I have discussed previously—our perceptions may be part of biologically-based evolution, but our rational frameworks are part of our *psyche*, the cultural part of our mind. That would invalidate any

Kantian explanation of objective reality as something culturally-biased. Although it is difficult to get rid of our intuition about reality, there definitely is a physical world out there, but it cannot be the way it is perceived (and rationalised) by human beings.

Also, as we saw before, quantum mechanics has demonstrated, perhaps not at our scale, that in reality, the observer, together with the object observed, both participate in the experience. According to the principles of the wave function, a particle can be in several locations at the same time (a superposition), but the moment it is observed, the particle has to break the superposition and end up in one random location. Reality behaves in a different way depending on the existence of an observer. Reality is also random. The other thing quantum mechanics has demonstrated is particle entanglement. Particles may be entangled and share information even when they are very far— even light years—apart. They do not need a physical mediator. There is change without physical connection.

So, quantum mechanics has challenged a principle that appeared unchallengeable. Again, as we stated before, the other issue with quantum mechanics is that it is not compatible with general relativity. Is a theory of quantum gravity needed to align those two? We'll see that perhaps they are actually as aligned as they're going to be.

~

SCIENCE KEEPS on quantifying and measuring objects. It keeps on analysing and trying to simplify things in order to understand them better. That is extremely useful in terms of technology, as it allows to predict with precision. However, some would say it is still necessary to reconcile the classic principle that a proposition can only be true without the bias caused by observation, with the principle that the location of an object may be changed only by being observed.

Science's need to quantify and measure only obstructs the study of consciousness, as *psyche* is cultural, and thus non-physical, intangible. So far, science has found that, with the means at its disposal alone, a full understanding of consciousness is impossible. *Psyche* has a deceptively subjective nature (because, being cultural, it's neither fully subjective nor fully objective), so attempting to use reductionist or objective methods has a negative impact on any research into it.

Intuition cannot quantify or measure the knowledge or information in our minds or in our culture, but it can infer/deduct that the knowledge in an individual mind has originated in the collective perception of reality. That collective perception is transmitted individually as each child is brought up.

Scientific analysis admits questioning, but it needs to be exhaustive in terms of imposing limits. Science divides, analyses, to understand. When things are expected to be a certain way, unfortunately, you come to wrong conclusions, or you ask questions that are not appropriate.

There are moments when explanation by analysis should come to an end. Neurologist Patrick House asks:

> *"How is it possible to even expect the word 'consciousness' to contain in itself the collapsed variation of billions of years of evolutionary differences? The concept, like the brain that contains it, has evolved."*

The concept has definitely evolved, and one of the aspects of evolution is that consciousness is not contained in the brain.

Psyche is a phenomenon generated outside of evolution (something that runs parallel to it). Perhaps you could call it an epiphenomenon: a by-product of evolution that keeps on occurring but has a separate reality. That way, AI is a derivative of a derivative. Nobody would attempt to describe intelligence of that nature as part of evolution, I hope.

Human consciousness and time have several qualities that science— especially physics —would seem unwilling or unable to deal with. Is it because they are non-physical? Perhaps it is impossible for science to deal with them. Perhaps, in any case, science tends to impoverish the qualities of consciousness and time in a systematic way because it does not have the elements to describe their reality in all its complexity. It falls short of the nature of both consciousness and time. Among these qualities, I refer more than anything to the intimate relationship that exists between them. The richness of that relationship is lost in a discipline that idealises, that needs to take the flesh out of physical reality and turn it into symbols to understand it better. But I have been using the term "qualities" and perhaps that is where the dilemma lies. Science rejects anything qualitative because it cannot measure it. Philosophy, psychology, and other humanistic disciplines do not need to do that.

We can say with complete certainty—and I will keep on repeating this—that without consciousness there is no time, and, without time there is no high consciousness, and no identity. Science has dealt with the consciousness-time relationship, but only marginally. Because before you do it, you need an exact definition of both—of time and of consciousness. That is impossible in scientific terms since objective reality is a tool probably created by religion and certainly used by science. And the perception of time is, by its nature, exclusively subjective, or subjectively shared. The perception of time—as we have already seen—resides only in the individual *psyche*, and is shared from different points in space.

Einstein combined time with space by conceiving of the existence of space-time. He did it in the field of physics and only as part of a scientific analysis. But he went much further: he said that there is not a single time, but an infinity of times. As many times as there are points in the universe. And that has consequences in our daily lives. That the clock on the table measures a different time than the clock that is on the floor really does not matter at our daily level. That there are infinite times does, especially if we want to define the nature of time to reach some conclusion, scientific or philosophical. That is to say that, by existing in consciousness, time has expressions as varied as there are individual consciousnesses and as the points in space in which those consciousnesses are, or those that can be imagined. Perhaps at some point a definition will be achieved that includes that many variables and makes sense.

Another of the coincidences between consciousness and time is the singular and at the same time multiple nature of both. We said that there is one time and infinite times. There would also seem to be a consciousness and what seem like infinite consciousnesses (the collective). High consciousness does not exist in the individual in isolation. As a linguist I know that consciousness is shared, as language is shared, which is the same thing that happens with time. That does not mean that there is an objective reality. Reality only seems to exist as the sum of the perceptions of individuals and their interactions with others.

∽

DARWIN'S PARADOX may lie in having come to the conclusion that the human being is part of the animal kingdom on the basis of a Western and Christian vision, which allowed him to study the animal kingdom from a privileged perspective. So, in *The Origin of Species*, Darwin applied scientific principles based on an objective reality to study himself, as part of the species.

The way I see it, honestly, trying to explain consciousness, time, or identity, from an objective reality based on physical particles, is like wanting to describe "Starry Night"—Van Gogh's painting—talking about the components of blue pigment. Poor Vincent was crazy and maybe that's why he shot himself. His madness would have been related to chemical reactions or nerve impulses within his brain, which included a tortured mind. But that same mind, which observed the night and imagined the picture, and the hand that painted it, had no simple relationship with either axons or synapses. Nothing to do with quantum particles. They had to do with an artistic expertise, creativity and sensibility that were part of his personality and genius. From my perspective, the possibility that that—complicated as it is—could be explained by talking about neuronal synapses is, to put it in scientific terms, less than 10^{-44}. And, if it could be done, how valid or relevant would it be?

∾

PARTICLES MANIFEST themselves in terms of probability. They could be in many places at the same instant. That uncertainty is part of quantum mechanics. The mathematical expression of the subatomic particle appearing in one place or another is called "wave function". The moment the particle is observed or measured, the uncertainty of the wave function disappears and there is certainty again: the particle is somewhere. Only one measurement is possible: only position or momentum can be measured. After that, the wave function collapses. Julian Barbour (*The End of Time*, 1999) says so, and I believe him. What Barbour says about measurement—and what quantum physicists face every day— is a puzzling aspect of reality:

> "It is not that before the measurement the particle does have a definite momentum and we simply do not know it. Instead, all momenta in the superposition are present as potentialities...".

If the subatomic particle is potentially anywhere, that means that there is no need for movement: at that level, there is no velocity, no space, and there is no time.

But the main problem with the collapse of the wave function is the fact that it happens when the particle is being observed and/or measured. To have an observer, or to have somebody measuring something, you need to have consciousness. Incidentally, the same thing happens with (general and special) relativity. The only difference between relativity and quantum mechanics resides in the fact that one needs an observer to have certainty (quantum mechanics) and, in the other one (relativity), certainty is a constant. We are all points in a block universe that is a static entity: past, present and future exist simultaneously, i.e., there is no time as such. And there is no objective reality in either. In both instances, all is relative to the observer.

$\approx$

THERE APPEARS to be a bit of a paradox, then: Western science needs to measure. It needs to measure in order to demonstrate reality objectively. To study and learn, we need to agree on certain concepts, and in order to agree, the reality of those concepts has to be demonstrated in terms of objective reality; i.e., we need objective reality to live in society. That brings certainty to our collective lives. That is one of the reasons we measure time (change), and space (distance), and velocity (change, move-ment, and space).

Human consciousness—we have seen—is a by-product of human society. Without the collective we would not have developed nor continue to develop *psyche* every generation. And time is a human construct, which we also need, and measure, to live in society. *Psyche* has created objective reality and time. It needs them. Science helps human society by demonstrating the existence of two things that do not exist (except in human minds), such as objective reality and time, but it cannot demonstrate how the entity that created those concepts—consciousness—emerged (or still emerges).

This brings me to the concept of objective reality, and my linguistic training tells me that there is something wrong here: the words are wrong. If society needs "objective" reality to live and grow and thrive and prosper, and if objective reality doesn't exist because it cannot be "objective"; if our senses cannot perceive reality the same way a dog or a tiger perceive reality, but science still needs it, then the term "objective" is inappropriate. Reality for us is what we perceive as reality and what we agree upon.

We do not need consensus about reality with all other living, sentient, creatures. We only need to prove and measure reality among human beings, among the collective. Then, perhaps, we need to think in terms of a "collective" reality. That is, the reality that all conscious, i.e., sentient and cognisant, humans share.

Thus, these two entities—collective reality and time —, although non-existent, unreal, are constructs consensually adopted by our collective consciousness because it makes sense to agree on their use.

Perhaps then, it would be important to recognise that both, collective reality and time, are only tools that allow science—with its limited, reductive, means—to understand the universe from a physical perspective but, thus far, that does not include the complexities of human consciousness.

CHAPTER 4

TWO DISCRETE LAYERS: SENTIENCE AND
SAPIENCE

*"*Human consciousness consists of two integrated but discrete layers:*
1) basic animal consciousness (nephesh)
2) high human consciousness (psyche).

* High consciousness is only acquired through
parental and collective upbringing. It is indi-
vidually transmitted. Its nature is cultural."

*H*ere we must make sure we understand
clearly what we are talking about. At
the end of the Introduction, I mention basic animal
consciousness and high consciousness (the latter, also
called sapience, which developed when our species
branched out of *Homo Erectus* and became *Homo
Sapiens*). I also use the words *nephesh* and *psyche*. I

believe these two terms are very useful, as they distinguish the two different layers of human consciousness.

We may love or fear other individuals of any species. We often love our parents, or relatives, or members of our collective, or pets. We do that instinctively. We may feel revulsion for something or even somebody. That may be what we call "a gut feeling". We know that other animals are capable of similar feelings. However, our *thoughts* about another person, or animal, may also result in feelings of love, or revulsion, or hatred, or fear. Those feelings come from our *human* consciousness because they are the result of thought.

Human beings can think about their own thought processes. That is metacognition.

From my perspective at least, not all sentient beings are conscious. Only human beings are conscious. Only human beings are capable of metacognition. Over many generations, we have developed sophisticated neural mechanisms. These accumulate, increasing with every generation; they grow individually but—through changes in our DNA—are integrated into our biological evolution: we can create new memories, and we can also remember events that took place many years before. We can do that at the individual level, but also socially. We are capable of operating within society; for instance, we can recognise faces in a way that is uniquely human. Other species can recognise some human faces to some extent, but never with the detail in which we recognise the facial features of human individuals.

A sentient being may have an experience, whereas, a rock—an object—cannot experience anything as far as we know. Neither of those cases involves consciousness. But the difference does not reside in the fact that one is an animal and the other one is a thing. A sponge, which is an animal, does not have senses. Something that is self-evident is that not all animal species have developed sentience. Some have no centralised nervous systems. Some are mere collections of cells.

Philosophers of science and scientists currently understand that any animal that has specialised neural structures and a central nervous system is a conscious being. To me, that animal is a sentient being, not a conscious being.

Consciousness has to involve *psyche*, the capacity to think and understand that the individual thinks, i.e., that it is a human being. That is why I prefer to use my own terms which, I believe, are clear, specific, and have a history. The way I see it, attempting to describe what it is like to be an individual of another species amounts to pure unfalsifiable speculation.

By *nephesh*—or sentience, if you like—I mean the capacity to experience sensations, or feelings, and emotions, to perceive objects and change. We share that capacity with other species. It is part and parcel of being alive. It does not include cognition, i.e., it does not include long-term memory, unlimited learning capacity, reasoning, sophisticated comprehension and knowledge. These are the skills we can use to make strategic decisions and to act on the basis of those decisions.

Nagel sees the difference between *nephesh* and *psyche* (not using those terms) as a difference clearly linked to a false assumption: that consciousness should be considered within the field of biology and—more specifically—that it should be taken as an evolutionary phenomenon:

> *"But to explain consciousness, as well as biological complexity, as a consequence of the natural order adds a whole new dimension of difficulty. I am setting aside outright dualism, which would abandon the hope for an integrated explanation. Indeed, substance dualism would imply that biology has no responsibility at all for the existence of minds. What interests me is the alternative hypothesis that biological evolution is responsible for the existence of conscious mental phenomena, but since those phenomena are not physically explainable, the usual view of evolution must be revised. It is not just a physical process."*

Agreed. And, after he discards substance dualism, he adds a footnote:

> *"But substance dualism would still leave biology with a huge problem similar to the one we are discussing: namely, why has physical evolution produced organisms of a kind capable of being occupied by and interacting with minds?"*.

I believe Nagel is talking about human minds. The human mind, cognition to be more specific, is a by-product of society, which is a by-product of evolution. In that respect, being a secondary derivative, the human mind is what I mention in the Preface to this book: a black swan phenomenon; something totally unexpected, improbable to say the least. Bayesian probabilities cannot account for such phenomena.

What Nagel is saying, among other things, is that the subdivision of disciplines within the biological sciences is something that needs to be revised. The problem with Nagel's analysis is that he includes both sentience and sapience as a conglomerate within the term "consciousness". Although not totally physical, sentience is a necessary product of biology, it is a quality that emerges directly from biology. Sentient animals have evolved using their senses and need them to survive. That is just the way it is; a brute fact of reality. Cognition, however, is not a further complication of sentience. It is a whole new development, an addition solely related to culture. Accounting for cognition on entirely reductive terms is what has taken science to the position it finds itself in right now. It is impossible to advance when the premise of the research is wrong. What is required now is for science to ask the right questions and start all over again.

Neuroscience repeats the same spin. Seth explains:

> *"It is widely agreed that experience arises from a physical basis, but we have no good explanation of why and how it so arises. Why should physical processing give rise to a rich inner life at all? It seems objectively unreasonable that it should, and yet it does."*

Everyone is saying that white is black.

It has to be true. *"It is widely agreed"* [by neuroscientists], then there is nothing else to discuss: *experience arises from a physical basis.* Of course, and *genes generate behaviour,* not the other way around. And one wonders why the whole field of neuroscience is going around in circles.

❦

CLEARLY, individuals of many other species possess self-awareness. However, to have a degree of identity you would need to live in a community and be aware that members of that community recognise you as yourself. We have an identity that remains with us throughout our lives, as we have long-term memories. We do not know if apes or other animals, like elephants, who live in communities, have the sense of identity that requires long-term memory. We know our "self" changes with time. In our fifties, our self has grown up, it is not the same self we had during our teens.

∼

I HAVE HAD FIRST-HAND experience of amnesia. The amnesia I experienced was transient, but there were episodes that recurred; therefore, what happened during those episodes remains very clear in my mind. Waking up from an amnesic episode there is the sensation of being an individual, of being somewhere, of existing, but there is no identity for a while. Self-awareness appears to me as a basic biological trait.

Identity appears as a derivative of self-awareness, but it has components that are definitely cultural, not biological. Identity overlaps self-awareness and requires language. It requires a name. After waking up, and after just existing for a while, we identify as somebody with a name and a memory, within a culture.

Here it would be useful to clarify, maybe, that self-awareness is biological in that sentient beings need to be self-aware, internally and externally. "Biological" does not necessarily mean "physical". Self-awareness and sentience are qualities of life. But they are not produced by the neurons or by the brain particles of animals. Maybe at the beginning of the evolutionary process, senses came to be because they were absolutely necessary for life. Perception of distance and size or shape of an object, perception of sound and colour, may not be necessary when the individual is a sponge. But the moment that individual moves and needs to feed, or needs to escape from predators, then those senses had to appear for the species to survive.

~

As I state above, "sapience" can be used to mean human consciousness. They are equivalent. In any case, for our purposes, let's use *psyche*.

Our minds, and those of other animals, can feel and process pain and suffering (*physical* pain and *emotional* suffering, that is). Our bodies, like those of other animals, are equipped to perceive pain because that is necessary for survival. Feelings like fear or apprehension are also necessary for the survival of any sentient being. We know that some animals, like us, feel grief and mourn the loss of close or loved individuals.

All those sensations, emotions, feelings and perceptions are known as *qualia*. These include how all sentient beings perceive objects, movement, change, colours, shapes, etc.

When I mention *nephesh*, that is what I mean: the basic abilities that sentient animals require to function within their environment. They are biologically generated and come with every individual the moment they become alive.

A recent article by Tim Bayne in *Big Think*, an online magazine, *"When do humans become conscious—in the womb or after birth?"* provides a clear idea of where the confusion lies. Humans cannot acquire consciousness in the womb. *Psyche* is cultural. We need to make that distinction fairly clear, as ambiguity appears to be the major problem in the study of human consciousness.

When Western scientists need to demonstrate that something is "objectively" real, they have to measure it and quantify it. Like all scientists, neuroscientists need to measure. They need pretty specific boundaries. But they have been going the wrong way about obtaining that definition they need. The dichotomy *nephesh/psyche* is pretty clear. And it should provide the boundary required in order to begin measuring. The difference is that *nephesh* is biological but has qualities that the individual needs to live. *Psyche* is not biological at all.

Bayne complains about the limitations of intuition:

> *"Unfortunately, going beyond intuition is challeng-*

> *ing. We cannot measure consciousness in the way*
> *in which we can measure temperature, pressure,*
> *and radiation. In the absence of a 'consciousness*
> *meter', we are forced to rely on inferences about*
> *infant consciousness based on the behavioral, cogni-*
> *tive and neural correlates of consciousness in*
> *adults."*

When does *psyche* begin at the individual level? Well, the answer is individual, but it certainly does not involve foetuses, as it is cultural, and foetuses have not been exposed to culture. It involves toddlers, depending on their degree of involvement in the culture that surrounds them. The development of *psyche* in an individual is directly proportional to their exposure to culture. Studying neural networks without discerning between cultural and non-cultural input will take scientists nowhere.

In his discussion of perceptual awareness in early infancy, Bayne asks himself:

> *"Is it really <u>human</u> consciousness?"*

It does not appear to be so, methinks... He continues:

> *"Infants presumably lack the capacity to stand back*
> *from their own experiences—to make them objects*
> *of critical reflection. Nor, presumably, do they have*
> *any sense of themselves as a continuing subject of*

awareness or as members of a community of conscious agents."

The comparison with adults is enlightening:

"Our awareness is not limited to the contents of immediate perception, but includes both memory and imagination — an awareness of how things were or how they might yet be."

We become fully human only when we realise that we share the world with the collective. Bayne ends the article on this note:

"When do we first acquire the capacity to subject our experiences to critical scrutiny? When do we first acquire the capacity to recollect the past or imagine the future? In some important sense, it is only when we have figured out the answers to these questions that we will know when 'human' consciousness first truly emerges".

He finds metacognition, memory and imagination. So near and yet so far. Neuroscience is reticent to recognise the presence of any cultural input.

$$\approx$$

* The logical conclusion is that there should be cortical and other centres in the brain, newer

> *than any centre that deals with strictly biolog-*
> *ical phenomena, where cultural developments*
> *are processed (i.e., Broca's and Wernicke's*
> *areas)."*

THIS IS one of the points I make in the Introduction.

Chimpanzees share 98% of our DNA. They are our closest relatives. They live in communities similar to the communities in which our ancestors lived, but they did not develop language or culture the way our ancestors did. Chimpanzees' brains are three times smaller than human brains. The cerebral cortex in humans is twice as big. The networks of their brain cells and our networks behave in different ways. How have our brains developed in such a different way?

The early stages of development of the brain—when cortex cells develop—appear to hold the answer to the problem. Recent studies by Mora-Bermúdez et al. have shown that human cortex cells are 50% slower to develop during an early stage called metaphase. Neuroscientists are now trying to explain why human brains are so much larger than chimpanzees' and if the slow development of these cells has anything to do with it. The answer cannot reside solely in the brain. The interaction between culture and the individual must have had some effect on the development of the human brain. Human offspring are much slower to develop than chimpanzee offspring.

Is there a link between brain development and culture? There appears to be a link between the expansion of our neocortex and our cognitive abilities (that are much more developed than those of chimpanzees). I believe the answer is quite apparent. Ours is an altricial species. We protect our children during the years of their development. We use our language to teach them how to behave and how to survive and thrive in society. This has happened for tens of thousands of years. And there is no chicken/egg dilemma here. Culture has had, and still has, an influence in the growth of our brain. Neuroscientists, however, continue studying the brain of humans and other creatures without a lot of research into the input from culture and the interaction culture/individual. Good luck to them.

~

WITHOUT TRYING to reinvent the wheel, and hoping for this not to appear didactic, I find that a clear metaphor to describe what I understand by *nephesh* is that it is the equivalent of the codified information that comes with the hardware of a laptop, for instance: the machine needs power to operate. Also, there are semiconductors and chips that direct information to different areas and produce results that end up being the functions of the computer. What organises all that activity is the operating system. We could say, then, that *nephesh* is comparable to the operating system of a personal computer, plus power. Without power and an operating system, the computer does not work. To be alive, any animal, including us, needs what the ancient Hebrews called *nephesh*: the breath of life.

This book is not about the emergence of *nephesh*. It is not about how life originated from non-life. That is for chemistry to solve. This book is about the origin of *psyche*.

Continuing with the analogy, every year there are new computer models. During the past few decades, we have witnessed the evolution of personal computers. Operating systems are upgraded periodically and personal computers evolve even further and require protection against malware, for instance. In general, computing becomes increasingly faster and, with the years, it includes more memory and sophistication.

What usually doesn't come with the personal computer are the different applications that make it really useful to us. Maybe a good metaphor for *psyche*, then, would be the apps that we use. These are acquired separately and interact with the operating system and the hardware.

The hardware, then, is physical. The operating system, which comes with it and is codified information, is necessary for the computer to function properly. As the name implies, it makes it operate. The operating system includes instructions (information) that activate certain physical areas of the computer. The other element required is power. Without power, the hardware doesn't work; with power, which is not physical either, it becomes alive. The software that has more specific uses, that is, the applications, make the personal computer a sophisticated, intelligent machine.

Our mind, our *psyche* (solely information), which is not physical, is the most sophisticated and intelligent piece of equipment mother nature has ever produced, and it has done it with the intervention of human culture. And, like any software application, we acquired it separately. How did that happen? I believe it is fairly clear.

CHAPTER 5

❧

HUMAN CONSCIOUSNESS

** Human consciousness is only acquired
through parental and collective upbringing. It
is culturally and individually transmitted. Its
nature is not biological but cultural.*

** Understanding the workings of high
consciousness cannot be arrived at through an
evolutionist study of basic— biological
—consciousness.*

hy are the above two points added at the end of the Introduction? Let's make sure we understand: the first one is quite evident. If we are discussing *psyche*, human consciousness—which includes cognition—, newborn babies lack that component. Newborn babies are obviously sentient, but have to learn language; they cannot express themselves except by crying. Parents, mostly mothers, but also the cultural environment, teach the child language. That is a process that lasts years. Once children have learnt how to articulate their thoughts, and other social skills, they can operate as individuals who are part of that social group, of that culture.

The second point stresses the fact that there is a big gap between sentience and sapience. Their nature is different. If our children could function once they have acquired physical skills like walking, running, swimming, and enough dexterity to create objects, they would probably be considered adults at twelve or ten years of age, maybe even less. But the age in which human individuals are considered adults in most countries is eighteen. Age of consent is usually sixteen. So, physical skills are not enough. Physical skills are biological. To function in the social environment the individual needs social and cultural skills. Understanding the former does not mean understanding the latter.

The study of consciousness has been plagued, from the beginning, by utter confusion. Because of the elusive nature of the subject matter, no agreement or clear definitions were ever reached on what the elements of consciousness are, and—to this day, as far as I know—there is no universal semantic consensus, or strategy concerning those elements.

Thomas Nagel, the philosopher, states in his book
"*Mind & Cosmos ...*":

> "*Certainly the mind-body problem is difficult
> enough that we should be suspicious of attempts to
> solve it with the concepts and methods developed to
> account for very different things. Instead, we should
> expect theoretical progress in this area to require a
> major conceptual revolution at least as radical as
> relativity theory, the introduction of electromag-
> netic fields into physics — or the original scientific
> revolution itself, which, because of its built-in
> restrictions, can't result in a 'theory of everything',
> but must be seen as a stage on the way to a more
> general form of understanding.*"

If we should suspect "*the concepts and methods*" of
physical Neo Darwinian science — which by definition
exclude the concept of mind — a non-physical entity,
then the solution must be found somewhere else.
From my perspective, the problem lies in the assumption that any evolutionist biological hypothesis must
include an evolutionary continuum. The solution to
that problem is that human evolution is not a continuum. The quantum leap, as we have discussed,
involves the introduction of language and culture,
and not just the introduction of language and culture
at a certain point, but as an ongoing development
that, since its beginnings, runs parallel — and is integrated into — biological evolution. That combination

results in an exponential growth of the human mind. Of course, the physical sciences will find that very difficult to accept. But accept it they will. All evolutionist explanations are bound to find a ceiling. Consciousness is not a process that can be approached solely from a physical, evolutionist viewpoint. It is one layer of an intertwined psychophysical process.

Now there are explanations like the one provided by IIT: all the universe is conscious. Is that scientific? Is that for real? Out of Cartesian dualism and into a new theism. A stab in the dark should not be considered science. Secular panpsychism is not science, it's a quasi-theological explanation. It does not render the universe intelligible. It is pure obscurantism.

One of Microsoft former executives published a story about three things he had learnt from Bill Gates: *"1) Dig for answers; 2) Smell the bull; 3) Synthesize from nothing."* The story applies to business, but it also applies to any problem that needs to be resolved, including scientific problems. It may sound simplistic but, in my experience, it applies to the "hard problem" of consciousness.

The origins of consciousness are lost in myth, religion and superstition. We will see, later, how that adds to the issue. The main problem, however—as it currently stands—is that the concept is opaque and nobody appears interested of clarifying what the term means.

Some say that being conscious is being able to have a subjective experience or being aware of something. Is being conscious, then, having experiences and awareness? Many animals have experiences, are self-aware, and are also aware of their surroundings. That is what I call *nephesh*: an individual of any species, that is alive and has senses. That is a sentient being. If we are interested in human consciousness—what I call *psyche* —we should restrict the terms "sapience" or "consciousness" to human beings, as that involves thought processes. To have an experience, a human individual (or an individual of other animal species) does not need to be conscious. They only need to be awake and aware of themselves and their surroundings.

Often our emotions and feelings come directly from our senses. When we hear sounds, or experience heat or cold; when we are scared, when we feel aggressive. These experiences may have nothing to do with our thoughts.

Sometimes thoughts and culture are involved, as our minds are an integrated whole. Stendahl, the renowned author, wrote about a visit to the *Basilica di Santa Croce*, in Florence where, after viewing Giorgio Vasari's monumental tomb of Michelangelo, he felt overwhelmed by the experience. He described "some sort of ecstasy" caused by the beauty of the monument and the fact that it was Michelangelo's tomb. There is a phenomenon often referred to as "Stendahl's syndrome", whereby sensitive individuals feel tremors, heart palpitations, perspiration, etc., when confronted with any form of sublime art. This can be visual or auditive.

The information in our minds may be integrated, but that does not mean that the system is totally homogeneous. I believe there is evidence that our minds are modular—I prefer the term "layered"; our different capacities appear to have come as responses to different needs of our species and at different times in the evolution of the species.

Without quite providing a full taxonomy of the human mind, modern philosophers Leibniz, Hume, and Locke articulated, to some degree, some of the concepts involved in the two-layered human consciousness.

Hume discussed *"impressions and ideas"*. The three philosophers drew a distinction between experience and thought. They do not appear to have discerned between biological and cultural inputs, though. Their distinctions are vague. Hume, again, thought that impressions and ideas were qualitatively identical, but that their difference resided in their intensity, i.e., a memory is not quite like the perception of an object. One is not as forceful as the other.

One interesting notion of Hume's is that there is nothing conscious that doesn't come from a sensation, i.e., that we must have a perception to be able to reflect on it: knowledge is derived from experience. But also, in these philosophers there is the concept of integration. We associate ideas.

Antonio Damasio comes pretty close to the same opinion: sentience is a minimalistic definition of consciousness. Normally, the latter should include memory, sapience, and metacognition.

∾

A RECENT ARTICLE IN *"NEUROSCIENCE NEWS"*, *"How young minds transition from gist to episodic memory"* provides a good example of how science is totally focused on the biological aspects of human consciousness and blatantly ignores any cultural input.

"Researchers discovered key molecular mechanisms that shift children's memory formation from general or "gist-like" to event-based or "episodic". This change in memory formation typically occurs between ages four and six.
The study identified that the maturation of inhibitory cells called parvalbumin-expressing (PV) interneurons, wrapped by a dense matrix called the perineuronal net, enables memory specificity and appropriately sized engrams.

> *This new understanding of memory development*
> *might unlock insights into conditions affecting the*
> *brain, such as autism spectrum disorder and*
> *concussion."*

The article mentions that important changes in memory formation occur between the ages of four and six. I am sure that is what happens at the biological level. Cells in the hippocampus mature and change memory traces from general to specific. There are molecular changes that result in these changes. This is all discovered through experiments with mice in laboratories. The connections between literacy and socialisation of children at that age are not mentioned by the researchers.

> *"The team notes that understanding this change,*
> *which generally occurs between four and six years*
> *old in children, may inform new insights in child*
> *development research and conditions which affect*
> *the brain, from autism spectrum disorder to*
> *concussion."*

There is mention of the growth of the perineuronal net that may trigger change in memory. But there is no mention of any interdisciplinary liaison between neuroscientists and educators or philosophers.

Memory traces in adults have 10 to 20 percent of neurons. In children, 20 to 40 percent. So, why does change occur? It all happens in the hippocampus, a centre for learning and memory. More cells are created of a type that constrains the size of the memory and allows for more specificity. No mention of literacy or of the richness of experience that comes with the socialisation of children in preschool/early school life.

Neuroscience has not worked out yet how the ability to form specific memories of events occurs in early childhood. The change from a fuzzy memory to a more episodic one, I am sure, helps in the passage from short to long-term memory.

We know that the hippocampus, which is part of the limbic system, is where memories are generated. We know that the limbic system is tightly connected to the prefrontal cortex, which is where all the major human changes have occurred in the brain. The hippocampus already has a basic memory system, appears to adapt to behavioural changes, and grows a more sophisticated, long-term, memory. That is, culture retrofits long-term memory onto the basic mammalian memory that is innate in human beings. This happens individually. Children are socialised and educated. Children are provided with information they need to store and retrieve in order to function within their culture. The hippocampus has evolved and continues to evolve genetically in a complicated process that involves the individual transmission of human culture. The sophistication of the human brain necessitates input from a continuous flow of cultural information. If culture is not provided, the development of the individual is thwarted.

I believe the main error lies in assuming that *psyche* must have somehow sparked chemically from matter; instead, *psyche* is the result of the cultural stage in the evolution of our species. The process is mentioned in more detail in the Chapter *Do we need a new paradigm for consciousness?* Particles and genes do not cause change, they follow it. Cultural and behavioural adaptation always precede physiological mutation. The way this continued mutation operates is through a combination of biology and culture.

Evolutionist research into human consciousness departing from basic creatures with a few molecules is also bound to find a biological ceiling. The involvement of culture in the process is a secondary evolutionary stage. Separate research is a must.

∽

WHEN KOCH and Chalmers made the famous bet in 1998, that by 2023 science would be able to explain consciousness, Koch believed in the possibility that science would be able to measure consciousness through the study of neural correlates. Chalmers, as a good philosopher, doubted the possibility of any association between brain activity and subjective experience. Saying that nothing has happened after that bet may be an exaggeration. There have been some advances and there have been different theories, some of them quite unfalsifiable and nonsensical, as we have described. But the question remains unanswered so far. Koch admitted defeat and the bet was renewed, for another twenty-five years I believe.

Part of the bet was an adversarial collaboration between researchers that investigated the neural basis of consciousness. There were experiments that attempted to decode brain signals to determine the markers of conscious perception. The results favoured IIT's assertion that consciousness arises from neuronal networks in the back of the brain. In any case, both theories had the shortcomings that were expected. More experiments are coming up.

To summarise much of what I discuss in this chapter, the ultimate question that Chalmers presented with his consciousness talk in 1994, highlighting the central mystery of human life: *"How do brain processes give rise to subjective experience?"* —or what is now known as the "Hard Problem of Consciousness"—has no answer because the question is based on the false premise that neurons in the brain generate subjective experience. It is inappropriate. In actual fact; it is a chicken/egg situation: subjective experience generates neurons in the brain. The question has no answer because it is the wrong question. The whole debate is based on the rejection of Cartesian dualism because of the current scientific bias towards physicalist monism. The mystery will persist until the scientific world decides to acknowledge that human consciousness has two layers that cannot be studied from a purely evolutionist point of view.

CHAPTER 6

IMAGINATION, CREATIVITY, LANGUAGE,
LONG-TERM MEMORY

** Imagination, creativity, language, long-term
memory, adventurousness, are exclusively
human traits acquired through high consciousness, i.e., through culture.*

** This confirms the Sapir-Whorf hypothesis
concerning linguistic relativity and falsifies
universalist claims concerning language, like
Chomsky's.*

uman beings have an amazing capacity
for learning. No other species comes
near that capacity. Long-term memory makes a big
difference. We remember and adapt to change. Israeli
academics Simona Ginsburg and Eva Jablonka

propose an evolutionary transition from minimal to high consciousness through *"unlimited associative learning"*. Although I personally cannot agree with the *"evolutionary"* transition part, I do believe that our unlimited associative learning is something distinctly human. Ginsburg and Jablonka list several conditions to reach the stage I call *psyche*: conceptualisation of objects; selective attention and active exclusion; integration through time; spontaneous activity; the existence of a goal; and the sense of being an individual separated from others and with a stable perspective over time. We fit the bill, indeed.

To do that, our brain uses a process called "prefrontal synthesis" or PFS. That means we can remember images of objects, or events, and we can apply those memories to imaginary situations. We synchronise these images of objects and events in our memories with possible objects and events that we imagine. Memory traces have to be intrinsically similar to the object or event remembered, but the combinations are endless. On that basis, we innovate. Imagination is involved in the process. We can think about possibilities of things that can happen in the future and we actually create those possibilities. Adventurousness is needed to create or to travel. Individuals of other species only migrate purposely within certain areas or as part of hunting tactics. Humans migrate long distances as part of a strategy.

Evolutionary naturalism is the only way physicalist science has to explain consciousness. It is a reaction to dualism. Evolutionism is viewed as the only secular alternative to theism. Actually, physicalism cannot cover thought, and theism provides no evidence of anything. Those two alternatives are an oversimplification, a nineteenth-century remnant of the dispute between Darwin and Wallace. Neither of those schools of thought can explain consciousness. But there are other alternatives. Science, however, is reluctant to accept any other paradigm. What it cannot explain physically does not appear to be real to it. However, human life is not purely physical. That is the core of the problem: dualism does not necessarily have to include a religious alternative.

Andrey Vyshedskiy, a neuroscientist from Boston University, states that we acquire PFS ontogenetically, i.e., as we grow up, and—he specifies—the way we grow up is a purely cultural phenomenon. Totally agreed.

Also, whenever a cultural phenomenon is involved, language has a lot to do with it. The more sophisticated language becomes, with the inclusion of recursiveness—when we use a subordinate clause, for instance—, the more sophisticated our mental processes become.

The acquisition of long-term memory, followed by PFS, must have resulted in the possibility of strategic hunting, among other things. Wolves can hunt in packs, but they only do it tactically. Humans began using strategy, long-term imagination, hence, traps.

Our species is gregarious. Without that quality and without language, it would have been impossible to develop the higher layer of our consciousness, the one that includes imagination, creativity and innovation.

Science has discovered areas of the brain in which memories are stored. The most important ones are the neocortex, the hippocampus and the amygdala. All these areas have specific functions, like the emotional content of the memory, and appear to have developed through culture. There are other areas, such as the cerebellum and the basal ganglia, which store implicit, short-term memories.

∼

IN A PREVIOUS BOOK, *Consciousness and Time - a New Approach*, I submit that human consciousness involves the two layers we discussed so far (*nephesh* and *psyche*). I also add that time is a human construct, again, as discussed above.

Saying that human consciousness involves two layers and that *psyche*, the exclusively human component, is culturally transmitted has consequences, implications and ramifications in all kinds of fields. *Inter alia*, it is a clear confirmation of the Sapir-Whorf Hypothesis in linguistics.

Let's revisit that hypothesis. In the early twentieth century, Edward Sapir was a teacher of anthropology and linguistics at Yale University. He thought that:

> *"Language is a purely human and non-instinctive method of communicating ideas, emotions and desires by means of a system of voluntarily produced symbols".*

Benjamin Lee Whorf, one of Sapir's students at Yale, believed language and culture were directly involved in the evolution of thought processes; he also believed that Western science ignored the differences caused by the phenomenon and that—consequently—Western scientific assumptions were unnecessarily narrow because they relied solely on a Western logical system.

Sapir and Whorf worked together on the hypothesis that the grammatical and verbal structure of a person's mother tongue influences that person's perception of the world. The idea, originally known as Linguistic Relativity, was fiercely rejected by universalists. It still is.

Coincidentally, that notion tallies with one of Whorf's studies that was also rejected as not a true reflection of reality: Whorf stated that the Hopi, a native American nation, lacked the concept of time in their language.

If time is a human construct—as I claim—that developed within human culture, or a device that we use to explain our long-term memory, if you like, it is possible for some cultures not to have developed the concept of time altogether, or to have developed a partial, or different, concept of time from the time we consider "normal" in the West.

We experience and interpret reality the way we do because we are predisposed by our language and by the way our culture perceives it.

Similarly, some Australian aboriginal languages include totally different ideas concerning pronouns, and times and modes of verbs.

The Pirahã, a fairly isolated Amazonian tribe, are probably a good example of linguistic relativity. The Pirahã language lacks cardinal numbers after "one" and "two", although they understand larger quantities; it has no colours except "light" and "dark" (it has other ways of explaining colours: "like-blood", for red); and it includes a system of pronouns—that can also become nouns—that is extremely difficult to understand by Westerners.

To give you an idea of the difficulty involved in understanding Pirahã grammar, their verbal system has a quantity of aspects: perfective (completed), imperfective (incomplete), telic (reaching a goal), atelic, repeated, and commencing; but they have very little transitivity. It appears fairly clear that the way the Pirahã language has developed is a reflection of the way they perceive the world. Languages and cultures develop according to the needs of a particular society in a given environment.

CHAPTER 7

∞

TIME

* Time is a human construct that exists only within high consciousness, through unlimited imagination (expectation) and long-term memory (which involves identity and collective perception).

* Without high consciousness there is only present and change.

J have mentioned research on paintings and notations conducted in hundreds of European caves, and on engravings of bones. Until recently, the depictions of animals were believed to be art. It has become evident that they were not art. They were mnemonic and notational devices. Our ancestors were observant enough and dexterous enough to reproduce images of deer, bison and other prey in ways that are understandable to us nowadays. They had not quite developed the idea of writing, but there is a germ of symbolism in their notations. Maybe I would not call them proto-writing, but the study found that frequently occurring signs, like dots, lines and "Ys", paired with figures of animals, were meant to carry meaning. The symbols signified months and seasons; they were part of a calendar beginning in spring and recording lunar months. The "Y" sign was indicative of parturition of the particular animal next to the notation. Did they have a concept of time? Maybe of change. I would guess that night and day were just never-ending cycles to them. Perhaps one day a chimpanzee will be able to tell us.

What is fascinating about that work is that it demonstrates, among other things, that human beings, since that early stage, had been measuring lunar months (time) for hunting purposes. That means that time (change and the repetition of seasons) had preoccupied humans for strategic purposes since the beginnings of human consciousness. The finding leads me to believe that there is a close association between the evolution of human consciousness and the concept of time.

Time has preoccupied human beings throughout history, and even during some prehistory, as explained above. But that preoccupation does not extend to other sentient animals.

As individuals, we learn about time from our parents, from society, from culture, from the language they use. Time is important to us as human beings. The way it develops in human minds is individual. Time is taught individually, again and again, to every child. It does not come as an instinct. We learn it.

I do not believe time disappears inside black holes. Time is not granular. Time may or may not bend in space, it depends on who's studying it, if anybody is. Physicists can speculate all they want, but time is only a human device, like numbers and identity. That is clear.

Time, like dreams, is an ethereal creation of *psyche*. The difference is that dreams are individual, whereas time is a tool used by the collective to have a clearer idea of when memories occurred, when expectations will occur, if they become a reality, and what causes what. As I say above, for all we know, our cave ancestors only thought about time in terms of seasons and the parturition of their prey because that was all they needed. Of course, they recognised night and day, probably as a never-ending cycle.

Maybe I should clarify what I mean when I claim the non-existence of time. Einstein clearly saw images of stars dying billions of years ago. He saw that in his own "time". That was happening then. The image was there. He realised, I am sure, that the only possibility for that was that, what he understood by "time" was something intimately linked to space. He then created the notion of space-time. From his perspective they were one and the same.

But I certainly believe what Aristotle said about it: "Time is the measure of change". Change occurs. Those stars dying were part of that change. For us, time is an infrastructure within which change occurs. It is the only way we have to explain causality or a succession of things and events at our level.

Time is the only way we have to explain long-term memory and long-term imagination. But it is cultural. And it is a creation, just like any number, or logarithm. These are concepts that we need to explain reality, like an ideal triangle explains the triangular shape of a mountain. The triangle is the idea that allows us to understand the shape, just like the infrastructure we call "time" helps understand causality, for instance.

The cultural nature of numbers is evidenced by their non-existence beyond the number of fingers in one hand in some languages. There are languages with numbers for "one", "two" and "many". Nothing else. Because that is all they need. One example I mentioned is the Pirahã language of the Amazon.

The fact that time is a cultural construct should not surprise us. It does, because we have grown accustomed to the concept. Time feels logical to us. Hard as it is to follow at first, once we realise that it is only an explanation for causality, and for our long-term memory and imagination, what becomes logical is that time is just the infrastructure we need to understand change.

I have stated on several occasions that we "*invented*" time. What I mean by that is that we have introduced the concept of an infrastructure where we can place change. Time, like our cardinal and ordinal numbers, like our colours, are cultural constructs. Of course, there is a colour for "green", but it does vary in hues and tones according to the culture, and it's probably nothing compared to the thousands of different greens hummingbirds can see.

Time is also cultural. That is evidenced by the fact that mechanical clocks were invented in the West, which is where they were most needed. Historically, different cultures have had different requirements and different ways of measuring time.

Currently, the fact that time is a construct of human consciousness is totally ignored in physics. Because of that, there is a proliferation of interesting articles that sometimes read like religious dogma. You can believe them, if you like, even if there is no possible way of proving or falsifying their contents. The following paragraphs are examples taken from an article by Sam Baron, of the Australian Catholic University:

" *A clock near a black hole will tick very slowly*

*compared to one on Earth. One year near a black
hole could mean 80 years on Earth, as you may
have seen illustrated in the movie 'Interstellar'."*

Well, the fact that a clock ticks very slowly near a
black hole may have all kinds of explanations. Will
we be able to demonstrate that empirically? Even if
the clock ticks slowly near the centre of our planet,
that does not mean there is an objective entity called
"time", although we can prove the part about the
slower ticking.

*" What about the past? This is where things get
truly interesting. A black hole bends time so much
that it can wrap back on itself.",* says Baron.

So, there is past in a black hole. Interesting. There is
also information, I am told. Maybe a vacuum cleaner,
as well.

*" There are three problems. First, you can only
travel into the black hole's past. That means that if
the black hole was created after the dinosaurs died
out, then you won't be able to go back far enough."*

You can travel into the past. Within a black hole. But
you cannot do it if the black hole was created after
dinosaurs became extinct. Really?

≈

SOMEWHERE IN THIS chapter I say that time is not granular. The concept that time is granular is introduced by Carlo Rovelli in his book *The Order of Time*. In a way, Rovelli is correct. I tend to disagree only because I believe time does not exist as such. When Rovelli declares that time is granular, what he is saying is that there is no continuity, there is no flow. The basic element of that apparent flow we understand as time is actually change, even at the subatomic level. Rovelli puts it very beautifully: *"Perhaps the rivers of ink that have been expended discussing the nature of the 'continuous' over the centuries, from Aristotle to Heidegger, have been wasted. Continuity is only a mathematical technique for approximating very finely grained things."* But he does it even more beautifully in the original Italian, when he continues: *"Il Buon Dio non ha disegnato il mondo con linee continue: lo ha trattegiato a puntini con mano leggera come faceva Seurat"*. He describes tiny dots, and those tiny dots are instances of change. He also mentions *"Il Buon Dio"* (*"The Good God"*). In a way that reminds us of Darwin in *The Origin of Species*, (or of Voltaire), Rovelli admits to a very scientific agnosticism. There may be a God, and he may be a pointillist, like Seurat was.

Rovelli and most other quantum physicists are rethinking time, as they study the deepest layers of reality. The time we are all used to does not appear to exist.

In his theory of general relativity, Albert Einstein also questioned the existence of time as a separate entity. He said it was bound to space. He called it space-time. Einstein said gravity warped space-time. In a way, his theory challenged objective reality as well. At that time, Newton's reality ceased to exist. Einstein presented a view that appeared not to be subjective. He called it "relative".

Once you question the existence of time as an objective entity, you are questioning science's entire paradigm concerning reality. If time does not exist for quantum mechanics—let's call it "change"—; and if time does not exist as such for general relativity—let's call it "space-change"—, one of the main obstacles towards a theory of quantum gravity would disappear I imagine.

Time is definitely a human construct. The inexistence of the concept before the appearance of *psyche* in humans proves it quite clearly. Human culture retrofitted the notion to explain long-term memory. And it also retrofitted it well beyond long-term memory, into the realm of imagination. When science contemplates the existence of the universe fourteen billion years ago, it is creating something that allows the human brain to interpret a past in terms of our planet rotating the Sun fourteen billion times. Of course, neither the planet Earth nor the Sun existed fourteen billion years ago. We can try to imagine fourteen billion years. I believe it is as useful as trying to imagine the existence of information in a black hole, or bidimensional particles in the event horizon of one. Good for physicists, not much good for anybody else.

Another interesting—and not coincidental—aspect of time is that, as part and parcel of reality, we need it to live in society. For our society to function properly, we need time, and we need a lot of precision in the way we measure it. When time started—when we began measuring change—those measurements were fairly loose.

In Europe, among the many devices that were used to measure time, they may have begun with the clepsydra— κλέπτειν (kleptein, to steal) and ὕδωρ (hidor, water)—from the Greek (a holed bowl that was filled with water and emptied within a certain period), which the Romans later adopted (and adapted) to measure the speeches of their senators.

Although we talk about the Greeks and its Greek name, the clepsydra is actually Egyptian. In 1500 BC, there was an inscription on the tomb of an official named Amenemhet that he had invented a type of clock that measured time during the night. The invention was an alabaster bowl with a hole and twelve marks inside bearing the names of the Egyptian months. The marks differentiated the length of day and night during the different seasons. Experiments have been carried out that suggest that the shape allowed a constant flow by which time was measured with a tolerance of between ten and fifteen minutes per night. Sufficient accuracy for the time.

In medieval monasteries, the day was divided into canonical hours—and in some of them it still is—; the hours were marked by ringing the bells and during those times they prayed or sang, or both. The liturgy of the hours, also applicable to all believers, began with the *matins*, more or less at dawn, and passed through the *lauds, prima, tercia, sexta* and *nona,* until reaching *vespers* and the *completas,* which was about nine o'clock, that is, the hour of rest.

A primitive alarm clock sometimes used was a candle with perpendicular nails at a certain height. As the candle burned and the wax melted, at the desired moment, the nails fell on a metal plate and made a noise that woke up the person.

Once humanity finally decided that the day had twenty-four hours, sundials appeared, and they were everywhere during Middle Ages, until mechanical clocks were first seen in the fourteenth century, on church towers and on the façades of town halls and other public buildings.

In 1582 the Gregorian calendar was introduced. Until then the Julian calendar had been used. The difference was that that year, from Thursday, October 4, it was passed to Friday, October 15, with the consequent loss of ten days. That corrected an error that the Julian calendar had had, in which the year was eleven minutes longer than it had to be.

We say that planet Earth is 4.5 billion years old. And we measure the time of its existence in years because it's the only way we can begin to understand that amount of time.

Something similar happens with distances measured in light years. A light year is the distance that light travels in a year, that is, 9,460,730,472.581 kilometres —that is, almost 9 and a half billion kilometres. No one, but no one, can imagine that distance. Although Einstein said he could imagine what it would be like to travel at the speed of light. A feat of his imagination, just like time is a human invention.

CHAPTER 8

❧

THE RETURN OF PANPSYCHISM

"Spinoza knew that all things long to persist in their being; the stone eternally wants to be a stone and the tiger a tiger. I shall remain in Borges... Thus, my life is a flight and I lose everything and everything belongs to oblivion..."
- *Jorge Luis Borges*

G iulio Tononi, an Italian psychiatrist and neuroscientist, David Chalmers, an Australian philosopher, and many researchers like them, believe that panpsychism might be the valid alternative to both materialism and dualism. There is a bit of *déjà vu* in the notion, as similar concepts were proposed many times in the past.

Giordano Bruno proposed that the whole universe was made of one substance and that consciousness was within that substance. Baruch Spinoza—as in the quotation by Borges—was of the idea that things had their own minds and that those minds we consubstantial with the mind of God. The Stoics, Plato and Aristotle had some kind of panpsychist notions, although panpsychism goes well beyond platonic essentialism. So, the concept is not new; what is new is that some quantum physics experts appear to confirm it to some extent.

Nowadays, Chalmers believes that *"some microphysical entities are conscious"*. Is panpsychism really a viable alternative to materialism and dualism? Galen Strawson (Real Materialism and Other Essays, 2010) believes so, and Phillip Goff (Galileo's Error, 2019) believes so. Of course, the whole concept has its detractors. Giants like Wittgenstein and Popper were very much against it.

The idea cannot be falsified, as far as I can tell; the onus to prove its veracity—however— should be on whoever wants to propose it as a valid theory.

Physics describes how matter behaves: for instance, it attracts, repulses, or resists acceleration. Nobody appears to know anything about its intrinsic nature. Matter is matter is matter.

What panpsychists claim is that the intrinsic nature of matter is consciousness. From my perspective, as discussed, the assertion is unfalsifiable. How can you look inside a subatomic particle and tell if it is conscious (or even sentient)? Human beings are conscious. And we can question how it is that we are conscious because we have a capacity: metacognition, not just because we are sentient. But, through MRIs and other brain-scanning devices, it is possible to ascertain if a brain is thinking or feeling, or communicating something, even if the individual cannot communicate normally. But, as far as I know, entanglement does not necessarily mean communication.

With all due respect to Spinoza and Bruno, and to some current scientists and philosophers, the fact that there have been some breakthroughs in physics does not mean that panpsychism is something that can be believed as the sole alternative to materialism and dualism. Quarks and electrons might have some form of sentience (or, you can call it, a germ of experience) —in any case, not *psyche*, not consciousness like the human one—but, either way, that is something that would require a lot of proving.

For sentience to exist now, there must have been a beginning, no doubt, and some kind of progression, evolutionary or not. We could talk about molecules becoming alive some millions of years ago on Earth. That would be a matter for biochemistry. On the other hand, to say that the universe has consciousness and that particles are even remotely conscious, requires a lot of faith, let alone research and experimentation. In the scientific world, things appear to be entering a "Holy Trinity" stage. There are mysteries that have to be believed, or else. To me, the whole thing sounds facile and smacks of desperation.

If we were to divide what is nowadays considered "consciousness" into *nephesh* and *psyche*, the whole problem would be reduced to: how did sentience begin? Evolutionist scientists like Ginsburg and Jablonka have no doubt advanced a great deal in that respect. There has been a lot of research in biochemistry as well.

But panpsychism appears bent on justifying the existence of human consciousness by reducing it to something common in the cosmos. We don't have to justify the uniqueness of our consciousness. As we said, it is a black swan event. It might very well be unique. And there would be nothing wrong or strange with its uniqueness. There are other, stranger theories and explanations of phenomena; we have a scientific community that advances solutions and notions such as a multiverse, string theory, or information within black holes.

In order to survive within their surroundings at a certain stage, living beings required sentience. They are called "sentient" beings. Our species is among them. Over and above that, we have developed *psyche* culturally, as a separate module, or layer, of our mind.

At the beginning of the 20th century, Eddington saw the actual "mystery" not as the mystery of consciousness but as the mystery of matter.

The only thing acceptable in terms of panpsychism, as well as something that favours the discoveries of quantum mechanics, is that it appears to falsify objective reality, a construct that has helped Western science for hundreds of years, but that now appears to have outlived its usefulness. Humans are not separate from reality; they are not superior to the rest of the animal world. Perhaps we should talk about "collective" reality. That would explain how humanity perceives nature in a certain way and how that perception is common to all human beings, but not to all sentient creatures.

Giulio Tononi proposed another concept: Integrated Information Theory (IIT). The theory states that a system becomes conscious when the information integrated in it reaches a threshold, that he called φ (phi). In 2015, Tononi, together with Christof Koch, co-authored a paper *"Consciousness: Here, There and Everywhere?"* which agrees with some of the principles of panpsychism. But panpsychism was a religion and what was thought to be everywhere was *'psyche'*, not in terms of mind, but the Greek concept of *"soul"*. It was like pantheism, which was a belief that God was part of everything in nature, i.e., a religion. Of course, panpsychism is unfalsifiable. Science is going into the stage Marcelo Seibler calls *"where science meets faith"*.

At the beginning of the chapter, I hint that there is some form of circularity to IIT. Robert Wright (*The Evolution of God*, 2009) says:

> *"On the one hand, I think gods arose as an illusion, and the subsequent history of the idea of god is, in some sense, the evolution of an illusion. On the other hand: (1) the story of this evolution itself points to the existence of something you can meaningfully call divinity; and (2) the 'illusion', in the course of evolving, has gotten streamlined in a way that moved it closer to plausibility. In both these senses, the illusion has gotten less and less illusory."*

What Wright states, if I understand it correctly, is that the fact that we created a hologram of God may in itself be God. If science is now saying that the universe is conscious, is it also saying that there is a meaning to it? Is science relying on hunches and intuitions to pave its way to unknown, non-physical regions?

That seems to be the current state of affairs in the study of consciousness. But then, if we accept that consciousness is an intrinsic property of matter, what we are indeed saying is that the universe is conscious.

What we said at the beginning of the chapter: some philosophers and scientists are proposing panpsychism as an alternative because they cannot see a way out of the dichotomy physicalism/dualism. A mini-consciousness would mean that if even inanimate objects have inner experiences, then, there is no division between subjective and objective. That, to me, appears as a facile solution, and on top of that, one that is not falsifiable. We would be taking a currently unknown element, such as consciousness, and placing it in the middle of the equation.

Physics—which so far works rather well, except for the wave function—is something that explains physical reality by abstracting it into mathematical formulas, i.e., non-material constructs. The irony is that, if we add a non-material element to that physical reality, i.e., consciousness, we would be adding a loop to it and complicating things no end.

Another panpsychism-related development is what scientists like Lee Smolin are proposing, which would end up making things even more difficult than they already are. From what I gather, Smolin attempts to include qualia in quantum mechanics and make the block-universe a time-related notion. It gets too involved to try to explain it here but, I believe, it gives us an idea of how misguided the scientific world is nowadays. Some scientists say that physics should not worry about things like consciousness; others—

like Carlo Rovelli—believe the answer lies in the way systems affect one another. We cannot separate subject from object. In any case, we appear to have come back to the Eastern view that objective reality does not exist. That idea makes sense for different reasons.

Panpsychism does not explain anything. From my perspective, panpsychism is a vain, half-hearted attempt at explaining both, consciousness and the wave function. The problem is that they are two separate phenomena.

What Rovelli says—that systems affect one another—appears to be closer to the truth. Noise requires a listener. Was there sound, or noise, when the initial singularity occurred? My answer is "No". There could be no noise because nobody could hear it. Is there sound when a tree falls in the woods and there is no sentient being to listen to it. My answer is still "No". You could place a recording device and then reproduce it. Yes, but the moment you reproduce it and somebody hears, there is a listener. Before then, there is no sound, no noise. Noise and listener are a system, like colour and vision. Is there colour without vision? Well, there needs to be some agency for colour to exist. That is, you need light, something emitting light, but you also need something or someone with vision to perceive colour (or shape for that matter). The bottom line is that consciousness cannot exist without agency. Matter is dead. Panpsychism advocates that no agency is needed. Schrödinger questioned whether the universe before consciousness would have been:

"... a play before empty benches, not existing for anybody, thus quite properly speaking not existing?".

Something similar happens with subatomic particles. They seem to be everywhere, but the instant they are observed, their position is fixed. The wave function theory implies there is motion (one of the variables being momentum, which is mass in motion). From the very basic understanding I have of it, the particles are not moving. Schrödinger and his cat. They could be anywhere, of course, but when they are observed or measured, their position is fixed. What happens is not that the particles decide to be somewhere. What happens is that, with the introduction of the agency (the observer) the system clicks. The cycle is closed.

Panpsychism, rather than providing a solution, denies that a system is needed. John Searle compares it to spreading a thin veneer of jam over the universe. Even that would require agency.

The other principle of quantum mechanics, that nobody appears to be using in order to explain anything, is that of entanglement. According to Schrödinger, again, it is:

"… the characteristic trait of quantum mechanics, the one that enforces its entire departure from classical lines of thought".

We still don't know how it works, that is, we know how whole systems work, but not the behaviour of their components. Particles can be paired. If a particle moves somewhere, and it has another particle paired to it, the other particle behaves in identical fashion. Entanglement can explain what Rovelli says: that systems affect one another. Entanglement could be used to attempt to explain noise and colour systems, and perception in general.

Attempting to reduce any form of sentience or consciousness, to mental particles in inanimate objects, however, seems absolutely nonsensical. Indeed, something that would need a whole lot of proving.

What Goff says in *"Galileo's Error"* is basically that panpsychism is a viable option. But then, that applies only if we are discussing immaterial souls. If we are thinking in terms of communication, information and *psyche*, we are not talking about souls in the ancient Greek or Christian sense of the word. One has to admit, it is a complicated subject. But once you recognise that there are other realities in nature that are not physical, the whole idea of panpsychism becomes ludicrous and facile.

On top of that, Goff appears to misunderstand materialism. All aspects of consciousness cannot be described by physical science. When he describes illusionism as a beautiful solution, all he is doing is adding more confusion to the subject of consciousness.

There is no doubt that consciousness and matter are interconnected; what is more, they are integrated, but in the human brain.

◇

THOMAS NAGEL, the philosopher, uses terms like "consciousness" (for "sentience") and "cognition" (for "consciousness"), which I would have used differently. However, he has a word that I find particularly interesting to describe *psyche*. The word is "process", and it includes humanity as the consciousness of the universe. I believe that is a much better way of explaining consciousness as an integral part of the whole cosmos. Humanity might be it:

> "The great cognitive shift is an expansion of
> consciousness from the perspectival form contained
> in the lives of particular creatures to an objective,
> world-encompassing form that exists both individu-
> ally and intersubjectively. It was originally a
> biological evolutionary process, and in our species it

has become a collective cultural process as well. Each of our lives is a part of the lengthy process of the universe gradually waking up and becoming aware of itself".

I am sure Carl Sagan would have agreed with that.

CHAPTER 9

$\mathscr{B}$

DO WE NEED A NEW PARADIGM FOR CONSCIOUSNESS?

*N*euroscience has been studying consciousness for some time now, since the second part of last century. Nowadays, research concentrates mainly on how wakefulness operates and, to some extent, how our mind deals with metacognition. Some of these studies involve indirect reference to the ideas I propose concerning discrete layers of consciousness but, in general, neuroscience is currently based on a monistic, physical, perspective. In the scientific world there is a clear materialistic trend reluctant to admit any social, cultural or inner subjective factors.

As we said, there are several current theories, the most influential ones among which are Tononi's Integrated Information Theory/ITT, and Baars/Dehaene's Global Workspace Theory. They have come up with some explanations of how the conscious experience operates. GWT is basically a description of metacognition. It does not explain the nature of the "hard" problem of consciousness. It does not explain what it feels like inside; the inner experience, which is where the problem resides. Tononi submits that consciousness is a fundamental aspect of the universe. Anything that possesses a certain minimum of "integrated information" experiences consciousness. As we know, the theory has been compared to panpsychism. Then, there is Friston's "Free Energy Principle", or FEP for short. ITT and FEP are opposite poles of the paradigm: FEP claims that "things exist", while IIT states that "consciousness exists". All of these theories, including IIT, are based on monist physicalism. Everything is physical, there appears to be nothing intangible.

~

THE MOST RECENT addition to neuroscience, a specialised discipline called "network neuroscience", has made distinct advances on the communication that takes place among the different function centres, or modules of the brain. There are modules devoted to hearing, motor movement and vision. Other modules have to do with attention, memory and introspection. These modules are independent but can operate in an integrated fashion.

Different modules control and integrate different lobes. The frontoparietal module, the one that controls the frontal, parietal and temporal lobes, appears to have developed relatively recently in terms of evolution, is extremely large in the human brain as opposed to the brains of other primates (many times larger), and is devoted to many cognitive tasks. It manages and synchronises other modules. The frontoparietal module is involved in how the individual makes choices and short-term memories; those choices, in turn, are developed into strategies and ethical behaviour.

All indications are, then, that the frontoparietal module, like the rest of the cortex, is not the product of a physical evolutionary continuum, but that it has appeared much later as the result of culture. It is a mostly cognitive module that integrates and unifies biological functions with cognition. It makes you human. It gives you your identity and allows you to retain it for life. I would call it the "cultural module". But science—especially neuroscience—does not currently accept that human consciousness operates independently of the physiological aspects of the brain.

A fact worth repeating again and again, is that human consciousness is not an emergent quality, it is the product of outside information that comes from the collective. It may result in biological or DNA changes, but it is not just another biological process; it is, above all, a cultural input that the individual shares with society.

What seems strange is that neuroscience studies the transformations that occur in the brain during childhood, through to the development of the adolescent brain, and until the individual becomes an adult; it also acknowledges that some of those changes occur when the individual learns new ideas and skills, but it does not discern between cultural and biological functions. For instance, it states that neural connection networks may exist in the womb, but not that those connections are only physical ones; any other conscious component is acquired during childhood.

A recent article by Bertolero and Bassett in *Scientific American*, for instance, provides some idea as to the mostly biological approach neuroscience takes concerning the development of human consciousness:

> *"... network neuroscientists have begun to ask why brain networks have taken their present form over tens of thousands of years. The areas identified as hubs are also the locations in the human brain that have expanded the most during evolution, making them up to 30 times the size they are in macaques. Larger brain hubs most likely permit greater integration of processing across modules and so support more complex computations. It is as if evolution increased the number of musicians in a section of the orchestra, fostering more intricate melodies."*

> *"... we need to know more about how personal*
> *genetics, early-life development and environment*
> *determine one's brain's structure and how that*
> *structure leads to functional capacities. Neuroscien-*
> *tists have some knowledge from the human genome*
> *about the structure that gives rise to functional*
> *networks but still need to learn precisely how this*
> *process occurs."*

The very words used by Bertolero and Bassett provide the answer: *"early-life development and environment"*. They do not mention "individual transmission", I would say, because the idea that children are not born with all the capacities and skills that adult humans have is a notion that has been negated by universalists like Noam Chomsky and his followers. The fact is that parents and family groups teach their children, and that individual cultures do influence the way our brains develop and the way we think.

Basically, the monistic approach to the study of consciousness has resulted in the total isolation of disciplines, like neuroscience, in their research, even if neuroscientists deny the fact. There is no explanation for collective consciousness. There is no attempt to explore cultural phenomena. Quantum mechanics' findings to do with wave function and entanglement have been misunderstood and blindly followed as the confirmation of a monistic paradigm. Neuroscience, philosophy and physics—with theories like Tononi's

—have come up with the misguided notion that consciousness is an integral component of all matter. The pendulum is totally on one side and there is a dogmatic reaction to anything else. People who reject these ideas have been called *"incredibly stupid"*. Nobody denies the interconnectedness of systems in the universe. Taking monism literally and claiming matter is conscious in order to explain things like entanglement—in physics—and the failure to find consciousness in the brain, are reactions that defy all logic.

$$\sim$$

APART FROM ANY advances in the location of functions within the brain, comprehensive theories of consciousness are, so far, empirically untestable and quite opaque. The subject remains misunderstood by physics, neuroscience and philosophy, and there is a lot of controversy surrounding the most basic aspects of consciousness.

Although neuroscience acknowledges that other processes, such as memory and social cognition are involved, so far, it appears to operate on the basis that consciousness mostly resides physically in certain areas of the brain. The cultural transmission of human consciousness on an individual basis has been sadly ignored so far.

～

"Eastern philosophy says there is no 'self'. Science agrees".

THIS IS the title of an article written by psychologist Chris Niebauer in the eMagazine *Big Think*; Niebauer claims there is agreement between neuroscience and Eastern philosophy concerning the inexistence of the self. The assertion is misguided and misleading in several ways. The article includes a lengthy explanation about the left hemisphere of the brain being totally in charge of consciousness. The left hemisphere, apparently, enjoys scams and traps; it tricks us into suffering:

"This false sense of self, which is often equated with the incessant internal dialogue, contributes significantly to human mental suffering."

The article leaves you with this almost paranoid feeling that you are trying to entrap yourself.

Among other things, the article claims:

"But, unlike our physical body, "[the 'self'], does not perceive itself as changing, ending (except, perhaps for atheists, in bodily death), or being influenced by anything other than itself."

Well, I'm not sure about the author, but my "self" knows it changes, it is influenced by the opinions of other people, and it knows it will end (and I am an agnostic). Perhaps the answer lies in the middle ground, but not monistically: perhaps that of Heraclitus, a monist, but with his river and the metaphor which states that, clearly, *everything* changes. The river will not be the same next time, but neither will be the man who will cross it a second time.

There is a self that can think, observe, assess, strategise. Human beings possess cognition for a reason. We need it, with our identity, and long-term memory, creativity and imagination, because we live (and thrive) in society. And we do have a "self" which is indeed the result of a thinking mind. Nobody is trying to trick us into anything.

Weird as it sounds, the article continues:

> " *The great success story of neuroscience has been in mapping the brain. We can point to the language center, the face processing center, and the center for understanding the emotions of others. Practically every function of the mind has been mapped to the brain with one important exception: the self. Perhaps this is because these other functions are stable and consistent, whereas the story of the self is hopelessly inventive with far less stability than is assumed.* "

Well, there is no centre for the "self" because the "self" is everywhere, especially in the cortex. Metacognition, as the name implies, overlaps everything else.

≈

ANOTHER RECENT ARTICLE—EQUALLY misleading —discusses *"access consciousness and phenomenal consciousness"*. The author decides that access consciousness (memories and information processing) can be explained by neuroscience, whereas phenomenal consciousness (the subjective experience) remains a mystery. The author claims that metaphors like the Cartesian theatre only confuse us; phenomenal consciousness is an illusion; consciousness is an emergent property that results from different parallel processes in the brain; phenomenal consciousness is not a distinct entity that requires explanation. Consciousness is an emergent property!

≈

IN THE OTHER ARTICLE, Niebauer rejects the existence of the self as an illusion but it does not provide any solution other than Buddhism. Unfortunately, articles like these are the result of the current state of neuroscience. There is magic, mystery and dogma. Opaqueness commands the field.

≈

IN A RECENT VIDEO, Lisa Feldman Barrett, Professor of Psychology at Northeastern University and neuroscientist, rejects the idea of the triune brain, first insinuated by Plato and later resuscitated by Carl Sagan in the twentieth century—a brain composed of evolutionary layers: a reptilian one (stem and cerebellum), a mammalian one (limbic system), and a human one (cerebral cortex). Feldman Barrett literally asks the questions:

> *"Why did brains evolve in the first place? What is the brain good for?".*

Her evolutionist answer is that neuroscience has traced the genes that helped form brain cells and discovered that the brain did not evolve in layers. According to Feldman Barrett there are no sedimentary layers. Her answer for the incredible development of the human component of the brain (thirty to one when compared with other mammals) is that the difference is the time it took to develop. This makes no sense. The answer is that:

> *"… brains evolved because, at one point, one animal ate another animal deliberately"*

and a game of predator and prey commenced. Again, no sense. When you have more sensory systems—she claims—there are lots of parts to co-ordinate and that requires a brain. She attempts to explain sentience to later reject the possibility of questioning *why* anything evolved. Feldman Barrett says we can discover *how* an organ evolves and what functions it has. Everything in the brain is related to metabolism and to the reproductive function. We have to pass our genes on. That is her whole explanation for the human brain; it is wholly physical. Of course, she is a psychologist but does not even once mention consciousness.

∽

OTHER ARTICLES, like one by Eddy Nahmias, of the Neuroscience Institute at Georgia State University, recently published in *Scientific American*, confirm the state of things in neuroscience. The author claims that:

> *"Increasingly, neuroscientists, psychologists and pundits"* ... *"suggest brain processing responsible for conscious thinking simply cannot count as free will. They often say that people who believe in free will must be 'dualists' who are convinced that the mind somehow exists as a nonphysical entity, separate from the brain."*...

> *"If most people are not committed to a dualist view about free will, then it is a mistake to tell them that free will is an illusion based on the scientific view that dualism is false. Why, then, do 'willusionists' believe the opposite? It may have to do with the*

current state of knowledge. Until neuroscience is able to explain consciousness— which will require a theory to explain how our mind is neither reducible to nor distinct from the workings of our brain—it is tempting to think, as the 'willusionists' seem to, that if the brain does it all, there is nothing left for the conscious mind to do."

If the scientific world is not sure as to how responsible individuals are of their own acts and omissions, how can our legal system function?

There is no denying that cognitive neuroscience has a long way to go. Apparently, what scientists claim is that most people think there is a control centre 'in the brain' but neuroscience cannot find it. Is there no single site where perception and decision-making take place? Neuroscience claims there is no such place. Perhaps what is wrong is the monist conception of consciousness? Is our mind solely located in the brain?

But we do have a 'self' and we do have free will. Otherwise, our whole society would collapse. So far, we appear to function within society. Why all this problem? Well, the scientific world does not accept that there is cultural component to consciousness. We are self-aware and we are aware of the existence of others in society.

Is there consciousness at conception? Is there a *psyche*? I repeat, the answer to those questions is that babies, when they are born, are not fully conscious. They are sentient. The cultural component of their human consciousness is gradually acquired during infancy, with language and, later on, with literacy and numeracy. Thought processes require time to become more sophisticated and refined. Children imitate, experiment, learn. Eventually they become adult individuals. The gradual appearance of identity in toddlers can be easily demonstrated.

Neuroscience, psychology, and philosophy appear to be at the stage when they will soon have to accept those facts.

∾

LET us see what kind of an alternative we can provide to current science (neuroscience, philosophy, psychology, and physics, both relative and quantum). To do that, perhaps we should recapitulate and try and see things from a clearer—perhaps historical—perspective. And maybe everything will fall into place.

The ancients tried to explain consciousness by way of a spirit that inhabited within us. The Hebrews, during the times of King Hezekiah, described the quality of being alive as something God imbued animals with, a "breath of life"("*nephesh*"). In Greece, some four hundred years later, Plato came up with the allegory of a chariot: humans have sensations and feelings, and every individual has a "charioteer" that directs everything else, the "self". Greeks had a word for that: "*psyche*".

Then came the Christian Bible, and translators used that *Koine* Greek word to describe the individual immortal soul. Saul of Tarsus (aka St Paul) had superimposed an immortal soul on the "*nephesh*" of the Hebrews.

At that point, the platonic idea of the charioteer ("*psyche*") became the soul that made humans different from animals, who only had "*nephesh*". Human beings had both, a body and a soul.

The amazing thing about human beings was that they were totally different from animals, they were superior and separate. God had told human beings they were above the rest of the world, especially above the rest of the animal kingdom. As a separate entity, they could study the rest of the world; everything outside a human being represented an "objective reality".

The Roman Empire adopted Christianity, and with it, the philosophy of an "objective reality". Western civilisation grew within that mentality. The world, the universe, could be studied as an object. The mind of a human being was the subject and the rest was the object.

There was a separation, a dualism. Material things were physical and concepts were spiritual. In the sixteenth century, Galileo Galilei, an astronomist and a physicist, among other interests of his (he was basically a polymath), proposed that science should concentrate on the physical aspects of reality. Science should study the movement of objects and should describe the laws that ruled those movements; mathematics could then describe those movements and those laws.

Philosophers like Descartes and Leibniz continued deciphering reality in different ways. Descartes proposed the dualism of mind and matter: he proposed that they were two different substances. Leibniz, did not accept the theory: he affirmed that there was only one substance in the world; a type of monism, whereby there was a special harmony between mind and body. But the problem with monism—we know—is that it leads to extreme notions like panpsychism, which do not make a lot of sense: a stone does not want to be, or remain, a stone. If it does, I would like a clear explanation of how that happens.

Isaac Newton—an alchemist, born a year after Galileo had died—took up the baton and continued with Galileo's ideas. Science was born. Phenomena could be studied and understood. There were rules, there were universal laws. The behaviour of matter was predictable.

In the nineteenth century, Charles Darwin—a biologist who had the opportunity to travel the world and study the different species of animals and plants —expounded on a theory of his illustrious grandfather, Erasmus Darwin, that species evolve. Charles Darwin's book, *The Origin of Species*, shook the world and introduced evolutionism and natural selection as an integral part of biology. Darwin's theory was based on random mutation, the genetic variation of species, inheritance of genes, selection of the most successful individuals and adaptation to the environment.

Russel Wallace, another naturalist who had come up with the idea of the evolution of species simultaneously, later claimed that natural selection could not provide a clear explanation for the advancement of the human species. He actually reneged evolutionist monism and returned to the religious views of his youth. He was a religious dualist.

Something Schrödinger said is very applicable here:

> *"In Darwin's theory, you just have to substitute 'mutations' for his 'slight accidental variations' (just as quantum theory substitutes 'quantum jump' for 'continuous transfer of energy'). In all other respects little change was necessary in Darwin's theory."*

Of course, religious views are currently rejected by most scientists. There is a wide consensus among researchers that the Neo Darwinian monist perspective is the correct one. There is no mind and matter. There is only matter that has evolved. Hence, neuroscience concentrates on the study of individual brains. But how can you explain consciousness emerging out of particles? Neuroscience explains brain activity, but cannot understand how conscious experience comes out of the brain.

Thomas Nagel, a Serbian-American philosopher and ethicist, questions the monist view that now prevails among scientists. He believes there are reasons to suspect that, using the current approach, neuroscience is going nowhere. Nagel is not a religious man; he claims, instead, that marvels occur in nature and that our species is one of them. Consciousness, he argues, is a fundamental aspect of nature. If a materialist or mechanistic view fails to explain that part of nature, it is misguided and will eventually fail. Coming from a fundamentally different perspective, I find that Nagel makes a strong point. I agree with his rejection of physicalism basically because physicalism cannot explain consciousness.

We have seen what philosopher David Chalmers called the "hard problem". The question is still there and it is still unanswered. I repeat: How can you explain consciousness emerging out of particles?

The answer is that consciousness does not emerge out of particles or genes. It is exactly the other way around: particles emerge because consciousness necessitates that emergence.

What happens is that particles and genes do not cause change, but follow it. There is a change, and genes adapt to it. Cultural and behavioural adaptations *always* precede physiological mutation. That, in turn, changes the DNA of the offspring. That is, changes in the behaviour of the individual generate genetic changes, it does not happen the other way around. What we do during our lives has a great impact on genes; that has been proved. Evolution presupposes a transition from non-sentient thing to a sentient being. And, of course, from a sentient being, at some point, evolution—or metaevolution, if you like— takes us to a conscious, thinking being.

Humanity involves the obsolescence of individual thought (our experiences and ideas generally die with us), which is accentuated by the finitude of the individual. However, the overlapping of generations partially solves the problem, as information is transmitted by parents to their children. The other important element in the transmission of information, knowledge and sapience is redundancy. It tends to counter or neutralise obsolescence. The coexistence of many individuals means that much information passes through to the next generation regardless of the disappearance of one individual. The collective usually knows more than the individual. As in physical reality, so it is in cultural reality: the whole is larger than the sum of the individual parts. This synergy means that information and knowledge are exponentially multiplied through the generations and geographically. Humanity grows in sophistication throughout the world. Information technology means even further growth.

The hypothesis that cultural transmission of information for thousands of generations necessitated the amazing growth of the human brain does not even need to be tested. It has been demonstrated already. Chalmers' "hard problem" is impossible to solve: it is in aporia. The question is the wrong question.

CHAPTER 10

FURTHER EVIDENCE OF THE
INTEGRATION OF CULTURE IN THE BRAIN

*I*n previous chapters we have discussed how behaviour generates DNA and not the other way around. Genetic changes are the result of behavioural changes. The large cerebral cortex in humans is not the result of biochemical changes which, in turn, gave us metacognition. Cultural and behavioural adaptation always precede physiological mutation.

~

RECENT STUDIES on cognition conducted at the Max Planck Institute for Human Cognitive and Brain Sciences, in Germany, have concluded that social isolation in the elderly leads to a decrease in the volume of the hippocampus and the thickness of the cerebral cortex. This, of course, affects memory formation, storage and retrieval and results in poorer cognitive performance, which increases the risk of Alzheimer's disease. The other side of the coin is that individuals who retain a strong social network have a better chance of preserving the structure of their brain and their cognitive function.

In the studies, loss of hippocampal grey matter and reduction of the thickness of brain cortex were directly associated with a decrease in social activity. Some of the participants who were not socially isolated at baseline but experienced social isolation at follow-up showed loss of volume of the hippocampus and cognitive decline.

A group of supportive family and friends, then, might contribute in the retention of brain health, and prevent the onset of Alzheimer's disease.

It's no accident, then, that a strong correlation exists between the thickness of the cerebral cortex and the inclusion of the individual in the collective; and this would not occur only in old age. The cognitive functions of the human brain develop as children grow and are socialised. These studies appear to add further evidence that the extraordinary growth of the human brain, compared to those of other mammals, is the result of social interaction among human individuals.

One of the areas that neuroscience does not appear to be exploring sufficiently—and not just evolutionarily — is the historical growth of the neocortex (an addition, the "new bark") through social interactivity and reciprocity. There is no need to repeat that the brain has grown from the centre outwards. Of the three areas of the brain, the more primitive ones are located in the centre of the brain. The neocortex has developed to respond to cultural interactions. That is, the neocortex and all brain areas related to cultural activities have grown exponentially because—and only because—there has been interactivity with the collective. Without that activity there would have been no growth.

∾

RESEARCHERS from the Department of Psychology and Neuroscience of Brigham Young University, in Utah, have recently proposed that the description of the brain as "triune" is incorrect because the brain operates as an integrated unit. They propose that the brain should be called "adaptive". The three key regions —they claim—the brainstem, the limbic system, and the cortex, do not function independently.

I don't believe anybody has ever proposed that they work as independent modules; it is fairly evident that they are integrated. Their claim is that

> "Specifically, emotion and cognition are interdependent and work together, the limbic system is not a purely emotional center nor are there purely

> *emotional circuits in the brain, and the cortex is not*
> *a purely cognitive center nor are there purely cogni-*
> *tive circuits in the brain."*

Again, nobody has proposed that the functions of the brain operate independently. New functions are retrofitted onto the old ones and operate as required by a new social reality.

We have seen in previous chapters that the hippocampus, which already has a basic memory system, appears to adapt to behavioural changes, and grows a more sophisticated, long-term, memory. The pre-human systems do not remain untouched by the evolution of the brain, the limbic system does not remain totally limbic. That is, culture retrofits long-term memory onto the basic mammalian memory that is innate in human beings. This happens individually. The new, developed, human brain is there, but children have to be taught how to use their innate capacity. The only thing innate at birth, as we saw, is brain capacity, not the information and skills, that have to be individually taught and developed.

Incidentally, the research at Brigham Young University that rejects the concept of the triune brain is purely evolutionary. In their paper they briefly mention social bonds as a key adaptation. They propose that:

> *"An important problem that early humans likely*

faced in surviving and reproducing was estab-
lishing cooperative relationships.".

By studying the brain monistically, as an individual entity—from a purely physical perspective— (i.e., no relationship with the collective), they go back in history and try to establish the individual reasons for its evolution, without any attempt at discovering current links with society.

The amazing growth of the neocortex did not—and does not—happen solely because early humans had to establish collective relationships. The convolutions of the brain keep on growing in cognitive complexity because the transmission of consciousness is an ongoing dualistic process. Cognition is a phenomenon that involves the integration of cultural and physical interactions, both interoceptive and exteroceptive as well. The neocortex, and the brain in general, grow accordingly.

∾

MYSTERIOUS SPIRAL SIGNALS have been discovered recently that apparently play an important role in the organisation or co-ordination of brain activity. It is possible that these vortices link different areas of the brain, which facilitates faster information processing. All of this happens in the outer cortex of the brain.

Traditionally, neuroscience has focused on the connections between neurons but has neglected research on social aspects of the cortex itself. The outer cortex, as we have discussed, is responsible for complex cognitive tasks. These studies indicate that neuroscientific research should shift towards the study of larger scale brain phenomena.

～

RECENT STUDIES in Wisconsin found that there was a higher concentration of myelin in language-related tracts in young children when the amount of adult-child interaction was intensive in the context of a rich linguistic environment. This contributed to the development of language circuitry. Children whose mothers had a higher level of education heard more adult words and developed speech at a faster rate than their peers in lower social and educational strata. Early brain development appeared to depend on the use of high-quality language during the early years. The link between a healthy brain/early language acquisition with a higher level of cognition in adulthood seems to be quite evident.

～

THERE ARE many more developments and they all point in the same direction.

CONCLUSION

"...If I have seen further, it is by
standing on the shoulders of giants."
- Isaac Newton

"No self is of itself alone. It has a long
chain of intellectual ancestors. The 'I'
is chained to ancestry by many
factors... This is not mere allegory,
but an eternal memory."
-Erwin Schrödinger

Newton's oft-cited metaphor possibly synthesises the spirit of this book. Perhaps science should specifically acknowledge that human consciousness reflects the linguistic and cultural aspects of the transmission of knowledge—in space and through the generations. Our neurons, our synapses, our individual brains, adapt, through an ongoing process, to the existence of a sophisticated society. That is, human consciousness includes an undeniable cultural component that has been seamlessly integrated—quasi-evolutionarily—into our human consciousness (and into its physical tool, the brain) in order to operate together with its other, innate, biological component: sentience.

Sentience, we discussed before, is just a brute fact of reality. Animals need it to survive. Humans have it, like all other animals. It is part of the evolution of creatures with a centralised nervous system. As animals developed, they needed to interact with their environment. The evolutionary result of those interactions is sentience. The physiological changes in those animals result from changes in their behaviour, not the other way around. Evolution explains how individual species adapt to their environment genetically through changes in their DNA.

∾

Here I have to digress. I need to recapitulate and try to point out, once again, the shortcomings of the current scientific establishment. This is what, from my humble perspective, they would need to correct. Their philosophical approach is erroneous. It needs to turn one-hundred-and-eighty degrees.

In his much-recommended book *Being You*, Anil Seth establishes quite clearly the paradigm currently favoured by neuroscience:

> *"My preferred philosophical position, and the default assumption of many neuroscientists is* <u>*phys-icalism*</u>*. This is the idea that the universe is made of physical stuff, and that conscious states are either identical to, or somehow emerge from, particular arrangements of this physical stuff. Some philosophers use the term* <u>*materialism*</u> *instead of physicalism, but for our purposes they can be treated synonymously."*

Well, there are clear problems with that approach. I discussed this before and I repeat it now: particles and genes do not cause change, nor can they generate it, they follow it. Cultural and behavioural adaptation always precede physiological mutation. The way in which brain particles operate cannot explain the social intangible. This sounds like common sense to me.

Seth continues:

> *"Sitting awkwardly in the middle [between physicalism and idealism]* <u>*dualists*</u> *like Descartes believe that consciousness (mind) and physical matter are separate substances or modes of existence, raising*

> *the tricky problem of how they ever interact. Nowa-*
> *days few philosophers or scientists would explicitly*
> *sign up for this view. But for many people, at least*
> *in the West, dualism remains beguiling. The seduc-*
> *tive intuition that conscious experiences seem non-*
> *physical encourages a 'naive dualism' where this*
> *'seeming' drives beliefs about how things actually*
> *are. As we'll see throughout this book, the way*
> *things seem is often a poor guide to how they actu-*
> *ally are".*

I take exception to absolutely everything in this para-graph. Especially, to the patronising attitude towards *'naïve dualism'*. According to Seth, then, *"conscious experiences <u>seem</u> non-physical"*. Really? Do they *'seem'* non-physical? How can anybody, in all seriousness, affirm that? I have not touched a conscious experience so far. I don't know anybody who has, either.

The prejudice is towards dualism, towards the notion that nature, especially human nature, may have a non-physical side to it. And that has to do with the fact that Descartes discussed anything non-physical as something 'spiritual'. Descartes was writing in the seventeenth century, though. Nowadays we can discuss abstractions and not presuppose that what we are discussing is anything 'spiritual'. Neuroscience appears to be rationalising an untenable position.

I have more bones to pick with Professor Seth, and they are not really with him, personally, but with the approach neuroscience has chosen. I'll go through one more of them.

According to Seth:

> *"An influential tradition, dating back at least as far Descartes in the seventeenth century, held that non-human animals lacked conscious selfhood because they did not have rational minds to guide their behaviour"... "I don't agree. In my view, consciousness has more to do with being alive than with being intelligent. We are conscious selves precisely because we are beast machines. I will make the case that experiences of being you, or of being me, emerge from the way the brain predicts and controls the internal state of the body. The essence of selfhood is neither a rational mind nor an immaterial soul. It is a deeply embodied biological process, a process that underpins the simple feeling of being alive that is the basis for all our experiences of self, indeed for any conscious experience at all. Being you is literally about your body".*

Seth discusses "selfhood". I don't know exactly what he means by that. The term does not discriminate between "identity" and "self-awareness". I presume he is talking about identity in human beings because he writes of *"being you"* or *"being me"*, and we are both human. If that is the case, he is sadly mistaken. "Being me" is being Rafael Pintos-López. It has nothing to do with an immaterial soul because I know I do not have one. It has everything to do with my rational mind, with my cognition. That is the identity I grew up into as a member of human society. I know that for a fact. And—as stated at the beginning of this book—I have had first-hand experience of amnesic episodes. Waking up from those episodes, pretty much like the instant one wakes up from deep sleep—but in a more profound, more prolonged way—one has the sensation of self-awareness without an identity. It is a very perplexing experience. One exists and one is an individual, but there is a lack of identity. I imagine—without any evidence for it—that that is what a tiger feels like. He has no identity (and he doesn't need one either). To feel human we need an identity, I can assure you. The alternative is mysterious, strange and scary.

~

Human individuals possess a unique ability. We can transmit complex information and we can do it collectively as well. We can pass an infinite amount of knowledge for countless generations. The acquisition of those skills is—I repeat—an ongoing process; and it is, somehow, built-into our consciousness.

There is a logic to it: in order to operate collectively, human societies require rules that have to be accepted by their members; human rules are intrinsically different from those of other animal groups. We have ethics, morals, and free will. What happens with human societies is that they do not operate following Darwinian principles. And Darwin knew that. Human societies are grounded on intellect, not biology.

Animal societies are basic and Darwinian; coexistence within them follows biological instincts. Human social structures require rules that have to account for —and predict—future events and contingencies as much as possible.

Life in animal societies is subject to all kinds of dangers and often follows the whims of alpha males or other powerful individuals (which can happen in totalitarian and primitive human societies as well). There is no trust, there is no strategy, change is unknown and totally unpredictable. Things occur day by day and individuals react instinctively.

The more advanced human societies operate with rules agreed upon and based on trust. Governments and other institutions rely on promises and have lifespans determined by means of collective or bilateral agreements.

At the beginning of this book, I mention that the emergence of human consciousness happens every day; it takes place within every individual. But the nature of human consciousness is such that it operates collectively as well. For instance, we have institutions such as money, which require the agreement of all the members of society. Different currencies are used to enable international trade. Very complex exchange operations involving huge amounts are conducted daily. The system, which has replaced barter and has been functioning for millennia, is evidently a cultural phenomenon. The value of a coin does not reside in the metal. Its value, like that of a note, is not physical, but institutional.

Basically, what happens is that we are individually taught how to superimpose and integrate language, thought, collective institutions and civilisation onto our biological reality.

Generations ago, somebody decided that gold was valuable and created coins; later, those coins were replaced by promissory notes or bank notes that represented those gold coins, as we discussed above. Eventually society came to accept notes and credit cards to buy real things on the basis of the promissory value of those pieces of paper or plastic. Now, even cell phones or electronic watches can pay for goods and services.

The most interesting aspect of collective consciousness is that the creation of an individual imagination can be understood by many. Also, members of human society can act collectively on the basis of institutions that are not physically real. Companies, associations, government agencies are among those institutions.

~

Approaching the study of consciousness has been traditionally considered difficult because of its constant recurrence, which involves a higher degree of sophistication in society, and complexity in the individual and in the collective—over and over again with each generation that passes. But the main problem for science has been in accepting the existence of some form of non-physical teleology. Teleology does not have to be associated with religion. Science takes for granted that there is no purpose to our existence. However, and regardless of the word "teleology", nature tends to repeat itself and become more complicated with every change.

In any case, if we use the term teleology as in "final causality", nature appears to be going somewhere. The term "evolution" itself involves the gradual development of living creatures. Humanity, especially, appears to be going somewhere. There is a direction of human society in terms of sophistication and complexity; that is difficult to ignore. It does not necessarily have to have any religious connotations. Thus far, for humanity, change has meant improvement.

Nagel explains it in these words:

> *"We recognize that evolution has given rise to multiple organisms that <u>have</u> a good, so that things can go well or badly for them, and that in some of those organisms there has appeared the additional capacity to aim consciously at their own good, and ultimately at what is good in itself. From a realist perspective this cannot be merely an accidental side effect of natural selection, and a teleological explanation satisfies this condition."*

Science has not rejected quantum mechanics because of its complexity. Complexity and recurrence appear to be accepted by science if they are quantitative and not qualitative. The scientific bias is obvious. What is also obvious is that there is a general trend in nature towards recurrence and complexity. Fractals are the clearest example. Fibonacci, Mandelbrot, and other mathematicians accepted the challenge—probably because it was measurable—but eventually found a formula.

Self-similarity is one of the qualities of consciousness. It exists at the individual and at the collective level in similar forms. Often, we use the term "microcosmos" when referring to small societies or individuals. The parallel mirrors that terrified and fascinated Borges are an example of self-replication to infinity. Labyrinths and gardens of forking paths appear to interest and terrify scientists as well. Perhaps the time has come to face the multifarious implications of the study of consciousness.

Above, I discussed how simple physical, biological, evolution cannot account for some principles that we apply in human societies; these have evolved, but only within human societies: we have *ethics* and *morals*, and we use those principles by means of our *free will*. By "ethics" I mean the standards that we normally apply to our actions at the individual level. To conduct ourselves ethically there are certain rules that we follow, which may vary among individuals, but which define what we perceive as right or wrong. The term "morals" has a pretty clear etymology, from the Latin *"mores maiorum"*, the habits of our elders; what we have been taught by previous generations concerning the standards of what is right and what is wrong, especially what is good or bad for society. Morality may vary with each generation, and from culture to culture, but there is definitely a benchmark provided by our ancestors. And, in free societies, we have the possibility to choose. Choice means that we can individually decide what is good or bad. Of course, if our choices do not meet the standards of our society, there will be social and/or legal consequences.

By adding a cultural layer to their understanding of nature, human beings create what Searle called "social reality". That cultural layer, that component of their consciousness, predates them, is integrated into their neurons, hubs, neural centres, synapses, etc., and is programmed to work seamlessly with their biological consciousness. Humans have access to the information and cumulative knowledge of all previous generations. They can see farther because they stand on the shoulders of giants.

ACKNOWLEDGMENTS

Inés Inés Inés Inés Inés Inés Inés Inés Inés Inés Inés
Inés Inés Inés Inés Inés Inés Inés Inés Inés Inés Inés
Inés Inés Inés Inés Inés Inés Inés Inés Inés Inés Inés
Inés Inés Inés Inés Inés Inés Inés Inés Inés Inés Inés
Inés Inés Inés Inés Inés Inés Inés Inés Inés Inés Inés
Inés Inés Inés Inés Inés Inés Inés Inés Inés Inés Inés
Inés Inés Inés Inés Inés Inés Inés Inés Inés Inés Inés
Inés Inés Inés Inés.

She read the draft many times, edited, commented,
discussed ideas, supported ideas.

Thank you.